AF479179

The Ocean and Coastal Conservation Guide 2005–2006

The Ocean and Coastal Conservation Guide 2005–2006

Edited by David Helvarg

Blue Frontier Campaign

Washington • Covelo • London

Library of Congress International Standard Serial Number

ISSN 1554-3137

Printed on recycled, acid-free paper 

Design by Maureen Gately

Manufactured in the United States of America

10 9 8 7 6 5 4 3 2 1

CONTENTS

ACKNOWLEDGMENTS

This guide began its evolution as research notes for the first edition of my book *Blue Frontier—Saving America's Living Seas*, an updated edition of which is available from Sierra Club Books. I was impressed by the number, energy, and diversity of the seaweed activists I encountered, as well as the organizations and institutions they were creating. Todd Baldwin and his colleagues at Island Press, a major source of information on the ecological state of our blue planet, came to share this enthusiasm to the point where they've agreed to publish the Blue Frontier Campaign's guide as a biennial book. Preparing the guide would not have been possible without the efforts of a number of key volunteers and interns at the campaign.

- Diane Williams, New York, is a mother of three, community organizer, and marine studies major. Diane managed to find time to research, check, and recheck much of the database that has evolved into this directory.
- Javier del Castillo is a surfing Texan with a love of the sea. A graduate of the University of Texas at Austin, he committed many weeks helping refine the directory and define many of the school, science, and marine park listings.
- Jon Christensen is a good friend and inveterate blue-water sailor from San Diego. A former financial manager with the county, he now teaches college part-time and works on estuary restoration. He also found time to help fact-check and review part of the directory.
- Jeff Oppenheimer, a graduate of UCSD, is Blue Frontier's California-based webmaster and one of the first people to begin organizing the guide as an online document and directory.
- Jean Logan, Blue Frontier's Program Director and fisherwoman extraordinaire. With great precision and an eye for detail, Jean spent the final weeks before this went to press recontacting hundreds of seaweed groups and individuals to make sure this product would be as accurate and up to date as possible.

Others who helped prepare the guide, offered useful tips, or reviewed certain sections include Michael Alguire and Carly Markowitz, who spent the summer of 2004 doing double duty helping prepare both a Blue Vision Conference and this guide; also Stephanie Yagos, Sebastian Howard, Christine Hoekenga, Dr. Gary Davis, Stuart Smits, Tony MacDonald, Jessica Brown, Buck Bagot, Cindy Zipf, Dawn Hamilton, Gerry Leape, and Lea Bonfiglio.

Blame any blatant errors or lapses on the editor, and he'll try to correct them in 2007.

INTRODUCTION

America is and always has been an oceanic society. From the Bering Sea land bridge to the Jamestown settlement to the processing lines of Ellis Island, we have been a tempest-tossed people, a saltwater people, a coastal people.

We have lived well on the abundance of our seas and coastlines from the earliest canoe tribes setting fish-traps along the Jersey shore to today's giant gantry crane operators unloading container ships at the Port of Long Beach.

As the Pew Oceans Commission report of 2003 and the U.S. Commission on Ocean Policy report of 2004 pointed out, America owes much of its wealth, bounty, and heritage to the blue in our red, white, and blue. It provides the oxygen we need to breathe, drives climate and weather, and brings rain to our farmers and food to our tables. It also provides recreation, transportation, protein, medicine, energy, security, and a sense of awe and wonder for the vast but fragile beauty of our blue-marble planet.

As the two commissions also point out, our living waters are today endangered by a cascading series of environmental threats from overfishing for the global seafood market to the nutrient pollution of our near-shore seas; from coastal sprawl that destroys the nurseries and filters of the seas; and from global warming that threatens our reefs, shores, and homes. In response to these threats, we need to develop and expand not only our scientific understanding of the seas but also an active and educated political constituency for the protection, exploration, and restoration of our living oceans.

That is why we've compiled this guide, as a kind of Blue Movement Directory, to identify and link citizen-activists, organizations, and institutions that, having gotten so much from our seas, are now ready to give something back.

The guide is divided into four sections. The first section identifies U.S. marine conservation organizations and provides descriptions and contact information by state. This is an easy way to find out who is active on what issues and in what part(s) of the country. The second section lists government agencies involved in coastal and marine oversight. It provides a way to navigate the confusing waters of marine governance in order to link particular issues to the agencies claiming legal responsibility over them. The third section lists major marine schools and science centers in the United States. It may be useful to potential students, activists, or other citizens seeking professional scientific or policy information and support. The final section lists national ocean sanctuaries and marine parks that can provide recreation and inspiration for all people who feel a connection to the sea around us.

Working together, dedicated watermen and waterwomen can begin to make a change, can create the first ripples in a rising tide of citizen action for the discovery and recovery of our Blue Frontier. We already know commonsense solutions can work, and that protecting our nation's watersheds, shorelines, and seas makes sense both morally and economically. Healthy, clean, and bountiful oceans will help ensure vibrant coastal communities and economies.

We hope this first biennial *Ocean and Coastal Conservation Guide* will provide a useful tool to the reader, while also helping to build a seaweed (marine grassroots) rebellion of creativity and solutions-oriented efforts across our land and from sea to shining sea.

David Helvarg
President
Blue Frontier Campaign (www.bluefront.org)

Blue Groups

I was visiting with the ranger in charge of a wildlife refuge on the coast of Florida where some of the last American crocodiles live in mangrove swamps and along abandoned canals of a failed development. The refuge is also home to endangered swallowtail butterflies, indigo snakes, and tree snails. Manatees, egrets, ospreys, tarpon, sharks, and rays congregate in the adjacent waters. I asked the ranger if he was involved with any ocean protection groups. "No," he said. "But when people ask me what environmental group they should contribute to, I say 'Planned Parenthood.'"

Fighting to protect our last wild blue places can lead some to a misanthropic view of the world—people are the problem. At the same time people also have the power to create solutions and find ways to live in harmony with nature and creation. American poet Henry David Thoreau reminds us that "Heaven is under our feet as well as over our heads." I'd just amend that to say that heaven is also under our flippers.

Across the United States people who care about our ocean planet are beginning to act on their beliefs. The number of blue groups is growing, along with public awareness about the state of our oceans. Basically five different types of blue organizations exist today:

- National and international activist organizations include the 150,000-member Ocean Conservancy (headed by a former Coast Guard admiral), which sponsors the National Beach Cleanup Day every year. The Surfrider Foundation is a chapter-based group of 40,000 surfers and other watermen and waterwomen who got fed up with oil and waste turning their ocean stoke into waterborne infections. Oceana, an international group, targets "dirty" fishing practices and threatens to sue the barnacles off polluting ocean liners. The Waterkeeper Alliance, with more than one hundred Keepers around the nation, targets polluters. The alliance grew out of the Riverkeepers who followed the flow into saltier water so that now they're also baykeepers, inlet keepers, and coastkeepers. These groups have local chapters or regional offices that work in your area and are easy to contact—particularly if your concerns are much larger than your own beach or bay, or if you are looking for an expert on a certain issue.
- Professionalized marine groups focus on just one thing and do it well. Seaweb helps the blue movement use media more effectively. It has spawned work on sustainable seafood, aquaculture, and aquarium conservation. Jacque Cousteau's grandson Philippe runs Earth Echo, which works on coral restoration and alternative ocean energies, while Titanic discoverer Bob Ballard brings the ocean into thousands of classrooms through his JASON project, using remote sensing underwater cameras. These groups can help your organization if you lack a particular kind of expertise. In another section of this guide you'll find listings of university and marine research centers that can also provide specialized expertise.
- Coalitions of blue groups across the nation, such as the Marine Fish Conservation Network, Restore America's Estuaries, Clean Water Network, and Ocean Wildlife Campaign, work on a range of issues from fisheries management reform to establishment of marine protected areas (wilderness parks in the sea). These groups often hold joint meetings and, like the Blue Frontier Campaign, aim to build bridges, alliances, and tactical coordination among marine activists and organizations.

- Local and regional blue groups include the New Jersey–based American Littoral Society, the Gulf Restoration Network, and Reef Relief that helps place tie-off buoys in the water so that dive-boats don't drop their anchors on live coral. The influential Chesapeake Bay Foundation in Maryland is fighting to restore America's largest estuary to at least 76 percent of its precolonial natural state. (Presently the organization rates the estuary at 26 percent.) Others are the venerable Save San Francisco Bay, the North Carolina Coastal Federation, and People for Puget Sound. Hundreds of local organizations also make their presence felt through constructive engagement with their fellow citizens. In Santa Cruz, California, for example, Save Our Shores (SOS), founded to protest offshore oil drilling, has evolved into a citizen-watchdog and resource for the Monterey Bay National Marine Sanctuary.
- Specific marine wildlife operations include the Pelagic Shark Foundation, Protect Our Wild Salmon, the Sea Otter Project, the Sea Turtle Restoration Project, and the Save the Manatee Club, which is cochaired by singer-songwriter Jimmy Buffett who notes that "each species is the spoke in a magic wheel: to lose one is to diminish the whole."

Ultimately, the Blue Movement includes many possible allies that you can call on—from local elected officials to ocean-dependent businesses, public health agencies, and recreational water users. In addition, a number of national organizations, including some major environmental groups such as the Sierra Club, the Natural Resources Defense Council, and Environmental Defense, now show an increased interest in the seas and offer members a way to be locally active while participating in campaigns aimed at changing national policies.

The following list of groups offers many choices and levels of activity to engage in. The list is organized by state so that you can find groups nearest to you working on your issues. It includes established institutions such as marine science centers, exceptional government agencies such as the California Coastal Commission, and conservation-oriented aquariums that offer teachers and students an opportunity to learn about ocean protection while also offering individuals a chance to volunteer. Neighbors are organizing hundreds of local groups to protect the coastal and ocean resources they most value, be it a recreational beach in Staten Island, New York, or traditional cultural resources for native Hawaiians and other Pacific Islanders. People have organized regionally around watersheds, treaties, and common waters at risk and, through their collective efforts, have won important victories for conservation and common sense.

Of course the real promise of what we call the "seaweed rebellion" is that all these groups—local, regional, and national—have now begun to work in concert. In July 2004 some 170 groups from twenty-five states and territories came together in Washington, D.C., for a three-day Blue Vision Conference that marked the beginning of a common vision and agenda for America's ocean and coastal movement. Building on that beginning, networks of waterwomen and watermen are today forging the kind of links that will help create a damp and salty uprising aimed at nothing less than the recovery of our maritime culture and heritage and the renewal of our journey home to the American sea. We hope you find the connections you need in this, the first citizens' guide to the blue movement.

ALABAMA

Alabama Coastal Foundation Inc.
www.alcoastalfoundation.org
PO Box 1760
Fairhope, AL 36533
ph: 251-990-6002
fax: 251-990-0041
joinacf@bellsouth.net
Improves and protects the quality of Alabama's coastal
resources by identifying and solving problems through education, cooperation, and participation.

Alabama Environmental Council
www.aeconline.ws
2717 7th Ave. S, #207
Birmingham, AL 35233
ph: 205-322-3126
fax: 205-324-3784
stateoffice@aeconline.ws
Strives to provide up-to-date information on Alabama's
coastal and terrestrial environment as well as links to other
environmental organizations. It is Alabama's oldest nonprofit environmental advocacy group.

Alabama Rivers Alliance
www.alabamarivers.org
2027 2nd Ave. N, Ste. A
Birmingham, AL 35203
ph: 205-322-6395
fax: 205-322-6397
alabamariv@alabamarivers.org
Works to unite Alabama's citizens to protect clean and
healthy waters through three programs that focus on
watershed protection, building watershed organizations,
and increasing awareness and understanding of the value
of watersheds to the state.

Dauphin Island Sea Lab
www.disl.org
101 Bienville Blvd.
Dauphin Island, AL 36528
ph: 251-861-2141
fax: 251-861-4646
gcrozier@disl.org
Provides educational programs, research and coastal zone
management, and a public aquarium focused solely on the
native ecosystems of the Mobile Bay estuary. DISL is home
to the Marine Environmental Sciences Consortium.

Mobile Bay Watch/Mobile Baykeeper
www.mobilebaywatch.org
5 North Jackson St.
Mobile, AL 36602
ph: 251-433-4229
fax: 251-432-8197
info@mobilebaywatch.org
Protects the quality of life for Mobile and Baldwin counties
through a number of programs that involve local citizens in
monitoring and reporting on water pollution, working to
improve air quality that impacts the area, protecting shorebirds and marine wildlife, and fighting a proposed liquid
natural gas facility.

The Nature Conservancy–Alabama Chapter Office
www.nature.org/Alabama
2100 1st Avenue North, Ste. 500
South Birmingham, AL 35203
ph: 205-251-1155
fax: 205-251-4444
lmayson@tnc.org
Purchases land and sets up conservation easements for critical habitats in Alabama to augment the nature preserves
of Alabama. The Nature Conservancy is a national organization headquartered in Arlington, Virginia.

**The Nature Conservancy–Alabama
Coastal Programs Office**
3280 Dolphin St., Ste. B-123
Mobile, AL 36606
ph: 251-473-4009
fax: 251-473-4809
nvickey@tnc.org

Sierra Club–Alabama Coastal Group
www.alabama.sierraclub.org/coastal.html
PO Box 852102
Mobile, AL 36685
ph: 251-540-2121
hodges@gulftel.com
Practices and promotes the responsible use of the earth's
ecosystems and resources including Alabama's coastal
lands and waters. The Sierra Club is America's oldest,
largest, and most influential grassroots environmental
group.

Weeks Bay Reserve Foundation
www.weeksbay.org
PO Box 731
Fairhope, AL 36533
ph: 251-990-5004
wernest@weeksbay.org
Supports the Weeks Bay Natural Estuarine Research
Reserve in its efforts to protect the pristine coastal area of
Baldwin County that feeds into Mobile Bay.

WildLaw
www.wildlaw.org
8116 Old Federal Rd., Ste. C
Montgomery, AL 36117
ph: 334-396-4729
fax: 334-396-9076
wildlaw@aol.com
Reviews and exposes violations of permitted industry and government entities in southern states, and ensures adherence to environmental laws. Has worked for the protection of Dauphin Island and other coastal regions.

Wolf Bay Watershed Watch Inc.
www.wolfbaywatch.org
PO Box 63
Elberta, AL 36530
ph: 251-987-5501
hammockhaven@gulftel.com
Protects and preserves the natural resources of the Wolf Bay Watershed. A grassroots citizen's advocacy organization.

ALASKA
Alaska Clean Water Alliance
PO Box 1441
Haines, AK 99827
ph: 907-766-2296
acwa1@seaknet.alaska.edu
Works to ensure that all activities in Alaska's rivers, lakes, estuaries, wetlands, and marine waters protect public health, support the use of the ecosystem for food harvesting, and foster environmental and economic stability. A research, education, and grassroots empowering organization dedicated to the conservation of the watersheds of Alaska.

Alaska Conservation Alliance
www.akvoice.org
PO Box 100660
Anchorage, AK 99510
ph: 907-258-6171
fax: 907-258-6177
unite@akvoice.org
Empowers citizens and organizations to participate effectively in the civic arena, and informs public officials, media, and the public about environmental, economic, and community-based issues. Among these are protecting salmon from impacts of logging and development and mitigating the impacts of global warming on the Arctic marine and coastal zone.

Alaska Conservation Voters
www.acvoters.org
PO Box 100660
Anchorage, AK 99510
ph: 907-258-6171
fax: 907-258-6177
info@acvoters.org
Works to protect Alaska's environment through public education and advocacy, and supports pro-conservation candidates for public office.

Alaska Longline Fisherman's Association
www.Afjournal.com
403 Lincoln St., Ste. 237
Sitka, AK 99835
ph: 907-747-3400
fax: 907-747-3462
Represents the men, women, and families dependent on Alaska's long-line industry and works to sustain it.

Alaska Marine Conservation Council
www.akmarine.org
PO Box 101145
Anchorage, AK 99510
ph: 907-277-5357
fax: 907-277-5975
amcc@alaska.net
Represents people who care about the health and future of Alaska's oceans and coastal communities. The community-based organization includes fisherman, subsistence harvesters, marine scientists, business owners, and families whose way of life, livelihoods and economies depend on healthy marine ecosystems.

Alaska Oceans Program
www.alaskaoceans.org
308 G St., Ste. 219
Anchorage, AK 99501
ph: 907-929-3553
fax: 907-929-1562
info@alaskaoceans.net
Shares concerns about the deterioration of the North Pacific Ocean environment and seeks solutions that promote conservation and sustainable fishery management policies.

Alaska Public Interest Research Group (PIRG)
www.akpirg.org
PO Box 101093
Anchorage, AK 99510
ph: 907-278-3661
fax: 907-278-9300
akpirg@gci.net
Educates citizens to enable them to participate in the political process, provides the public with practical and cost-efficient ways to work with government and the private sector, and encourages and provides information to grassroots efforts that advocate for the public interest. This is a nonpartisan, statewide, nonprofit organization that is over twenty years old and has about 3,000 members.

Alaska SeaLife Center
www.alaskasealife.org
301 Railway Ave.
PO Box 1329
Seward, AK 99664
ph: 800-224-2525
fax: 907-224-6320
visit@alaskasealife.org
Promotes research, rehabilitation, and public education to understand and maintain the integrity of marine ecosystems in Alaska. The nonprofit organization's research and rehabilitation facilities and naturalistic exhibits immerse visitors in the dynamic marine ecosystems of Alaska.

Anchorage Waterways Council
www.anchoragecreeks.org
PO Box 241774
Anchorage, AK 99524-1774
ph: 907-277-9287
fax: 907-277-9207
awc@alaska.net
Promotes the value of Anchorage's waterways and related habitats. The group organized itself following the first creek cleanup effort in 1984.

Campaign to Safeguard America's Waters
www.earthisland.org/project/viewProject.cfm?subSiteID=6
PO Box 956
Haines, AK 99827
ph: 907-766-3005
fax: 907-766-2921
gershon@aptalaska.net
Seeks to close loopholes in federal and state water pollution regulations that allow millions of gallons of polluted wastes to be dumped into public waters every day. This group is a national project of the Earth Island Institute.

Center for Alaskan Coastal Studies
www.akcoastalstudies.org
PO Box 2225
Homer, AK 99603
ph: 907-235-6667
cacs@xyz.net
Fosters responsible interactions with natural surroundings and generates knowledge of the marine and coastal ecosystems of Kachemak Bay through education, research, and stewardship programs.

Coastal Coalition
PO Box 231293
Anchorage, AK 99523-1293
ph: 907-333-3381
Seeks to develop coastal policy and a responsible plan for oil transportation, OCS oil development, and marine conservation. This private coastal coalition group formed in response to the *Valdez* accident.

Cook Inlet Keeper
www.inletkeeper.org
Box 3269
Homer, AK 99603
ph: 907-235-4068
fax: 907-235-4069
keeper@inletkeeper.org
Seeks protection of Alaska's Cook Inlet watershed and the life it sustains. The organization is a citizen-based nonprofit group and member of the Waterkeeper Alliance.

Earthjustice Legal Defense Fund–Alaska
www.earthjustice.org/regional/juneau
325 Fourth St.
Juneau, AK 99801-1145
ph: 907-586-2751
fax: 907-463-5891
eajusak@earthjustice.org
Seeks to protect the natural attributes of Alaska. Since 1978, attorneys representing a diverse group of clients have brought litigation to defend natural resources throughout the state.

Friends of Southeast's Future
www.ptialaska.net/~snep
PO Box 212
Sitka, AK 99835
johnson@ptialaska.net
Works to protect the Tongass National Forest surrounding Sitka from large-scale clear-cutting of the remaining old growth forest that protects local lands and waters. This is a grassroots activist group.

Greenpeace–Alaska
www.GreenpeaceUSA.org
125 Christensen Dr., Ste. 2
Anchorage, AK 99501
ph: 800-326-0959
info@wdcgreenpeace.org
Uses nonviolent direct action and creative communication to expose global environmental problems and to promote solutions that are essential to a green (blue) and peaceful future. Greenpeace is an international independent membership organization whose goal is to ensure the ability of the Earth to nurture life in all its diversity. Its present campaigns include efforts to protect ocean seamounts and eliminate destructive bottom-trawl fishing.

Kenai Watershed Forum
www.kenaiwatershed.org
PO Box 2937
Soldotna, AK 99669
ph: 907-260-5449
fax: 907-260-5412
robert@kenaiwatershed.org
Works to maintain the health of the Kenai Peninsula watersheds and ensure the quality of life for future generations.

Ocean Champions

www.oceanchampions.org
645 G St., Ste. 100, #617
Anchorage, AK 99501
ph: 907-258-8935
fax: 907-222-6202
jack@oceanchampions.org
Works to support proven leaders and elect new candidates
to the U.S. Congress who will advocate for ocean conservation. Ocean Champions is a nonpartisan organization.

Ocean Conservancy–Alaska Regional Office

www.oceanconservancy.org/dynamic/aboutus/offices/
alaska/alaska.htm
425 G St., Ste. 400
Anchorage, AK 99501
ph: 907-258-9922
fax: 907-258-9933
Uses science-based advocacy, research, and public education to inform, inspire, and empower people to speak and
act for the oceans. The Ocean Conservancy is a national
organization dedicated to protecting ocean ecosystems. The
organization has a national office in Washington, D.C., and
ten regional offices, including one in Alaska. Its four major
programs are coral reef protection, debris monitoring,
stormwater runoff and pollution, and coastal cleanup.

Prince William Sound Conservation Alliance/
The Eyak Preservation Council

www.redzone.org
PO Box 460
Curdova, AK 99574
ph: 907-424-5890
fax: 907-424-5891
eyak@redzone.org
Concentrates on all environmental issues facing Prince
William Sound, including the impact of cruise ships, oil
tankers, fishery and wildlife issues on the Sound's resources
and people.

Prince William Soundkeeper

www.pwsoundkeeper.org
PO Box 2832
Valdez, AK 99686
ph: 877-686-3533
info@pwsoundkeeper.org
Seeks to protect Alaska's Prince William Sound and the life
it sustains. This organization is a citizen-based nonprofit
group.

Prince William Sound Science and Technology Center

www.pwssc.gen.ak.us
PO Box 705
Cordova, AK 99574
ph: 907-424-5800
pwfsc@pwssc.gen.ak.us
Promotes the goal of maintaining long-term, self-regulating
biodiversity, productivity, and sustainable use of renewable
resources through ecological monitoring and educating
youth and the general public about the critical interdependence of the biology and regional economies of Alaska.

Sierra Club–Alaska, Anchorage

http://alaska.sierraclub.org
333 West 4th Ave., Ste. 307
Anchorage, AK 99501-2341
ph: 907-276-4048
fax: 907-258-6807
nw-ak.field@sierraclub.org
Uses lawful means to practice and promote the responsible
use of the earth's ecosystems and resources, and to educate and enlist humanity to protect and restore the quality
of the natural and human environment. The Sierra Club is
America's oldest, largest, and most influential grassroots
environmental group. Works to stop logging, drilling, and
other marine impacts in coastal zones, including the
Southeast and Arctic regions of the state.

Sitka Conservation Society

www.sitkawild.org
PO Box 6533
Sitka, AK 99835
ph: 907-747-7509
fax: 907-747-6105
info@sitkawild.org
Works to protect the natural environment of the Tongass
forest and waters of southeastern Alaska by educating citizens on environmental issues and helping them participate
in the public process to advocate responsible stewardship
of natural resources.

Southeast Alaska Conservation Council

www.seacc.org
419 6th St., #200
Juneau, AK 99801
ph: 907-586-6942
fax: 907-463-3312
info@seacc.org
Safeguards the integrity of southeastern Alaska's unsur-
passed natural environment while providing for the sustainable use of our region's natural resources. Among its campaigns, SACC is working to conserve Berners Bay, called the
"Serengeti of Southeast Alaska," and improve the state's
marine ferry service as an alternative to road building.

**Surfrider Foundation–
Wildcoast Organizing Committee**
425 G St., Ste. 400
Anchorage, AK 99501
ph: 907-258-9922
orca@acsalaska.net
Uses conservation, activism, research, and education to
encourage people to protect and enjoy the world's waves
and beaches. The Surfrider Foundation is a nonprofit environmental organization whose membership consists mainly
of surfers. Its scope has expanded to include protecting
wetlands, bird life, and beaches as well as surf spots.

Tongass Conservation Society
PO Box 23377
Ketchikan, AK 99901
ph: 907-225-5827
fax: 907-225-5827
TCS works to preserve the biodiversity of the many island
forests of the Tongass through protection of habitat from
pressures of industrial exploitation. Campaigns to protect
the unique and inspiring Misty Fiords, Ward Cove, and
other marine waters from the impacts of poorly regulated
tourism, pollution, and other activities.

Trustees for Alaska
www.trustees.org
1026 West 4th Ave., Ste. 201
Anchorage, AK 99501
ph: 907-276-4244
fax: 907-276-7110
ecolaw@trustees.org
Provides legal council to sustain and protect Alaska's natural environment, including coastal bays, salmon rivers and
marine wildlife.

ARIZONA
**Intercultural Center for the Study of Deserts and
Oceans, Inc. (CEDO Intercultural)**
Cedointercultural.org
PO Box 44208
Tucson, AZ 85733-4208
ph: 520-320-5473
info@cedointercultural.org
Seeks to advance and share knowledge about the Upper
Gulf of California and surrounding Sonoran Desert and to
promote conservation and sustainable use of its resources.
For more than twenty years this collaborative effort
between U.S. and Mexican nonprofit groups has main-
tained a natural history-marine science center and biologi-
cal field station near Puerto Penasco, Sonora, Mexico,
approximately 215 miles southwest of Tucson and Phoenix.
Here eco-tours and educational programs are conducted for
U.S. and Mexican students and interested adults.

CALIFORNIA
Adopt-a-Waterway
www.adopt-a-waterway.com
919 Manhattan Ave., Ste. 100
Manhattan Beach, CA 90266
ph: 310-374-3812
fax: 800-890-2213
heather@eccmedia.com
Seeks to improve the quality of the nation's oceans, rivers,
streams, and waterways. This group is a part of a national
organization.

Algalita Marine Research Foundation
www.algalita.org
148 Marina Dr.
Long Beach, CA 90803
ph: 562-598-4889
fax: 562-598-0712
info@algalita.org
Engages in innovative research, education, and restoration
of the marine environment with the help of its chartered
research vessel, the *Alguita*. The California-based nonprofit
environmental organization has been a major force in identifying the problem of plastic pollution in the open seas.

**American Cetacean Society–
Channel Islands Chapter**
www.acschannelislands.org
PO Box 3924
Santa Barbara, CA 93130
acf_sb_v@hotmail.com
Engages in educational, conservational, and scientific pur-
suits for the purpose of expanding scientific knowledge
about whales, dolphins, porpoises, and related creatures.

American Cetacean Society–Los Angeles Chapter
www.acs-la.org
PO Box 1208
San Pedro, CA 90733-1208
ph: 310-548-8500
acsla@acs-la.org

American Cetacean Society–Monterey Chapter
www.starrsites.com/acsmb
PO Box HE
Pacific Grove, CA 93950
seastham@monterey.K12.ca.us

American Cetacean Society–Orange County Chapter
www.acsonline.org/orangecounty
C/O 6241 Warner Ave., #203
Huntington Beach, CA 92647
ph: 714-375-1192
contact_oc@acsonline.org

American Cetacean Society–San Francisco Chapter
www.acs-sfbay.org
PO Box 784
Pacifica, CA 94044
ph: 650-738-4250
shari@acs-sfbay.org

Aquarium of the Bay
www.aquariumofthebay.com
Pier 39
Embarcadero at Beach St.
San Francisco, CA 94133
ph: 415-623-5300
fax: 415-623-5324
catherineb@aquariumofthebay.com
Seeks to preserve the rich and diverse aquatic life of San
Francisco Bay and its surrounding waters. This unique edu-
cational and entertainment facility is located on the San
Francisco waterfront.

Aquarium of the Pacific
www.aquariumofpacific.org
Pacific Aquatic life
100 Aquarium Way
Long Beach, CA 90802
ph: 562-590-3100
fax: 562-951-1629
aquariumofpacific@lbaop.org
Seeks to instill a sense of wonder, respect, and stewardship
for the Pacific Ocean and its inhabitants and ecosystems. This
facility is one of the largest aquariums in the United States.

Aquatic Adventures Science Education Foundation
www.aquaticadventures.org
1010 Santa Clara Point
San Diego, CA 92109
ph: 858-488-3849
fax: 858-488-4265
sharafisler@aquaticadventures.org
Provides educational programs that connect underserved
youth to science, inspire environmental action, and increase
exposure to marine habitats.

Arc Ecology
www.arcecology.org
833 Market St., Ste. 1104
San Francisco, CA 94103
ph: 415-495-1786
fax: 415-495-1787

arc@igc.org
Promotes peace, environmental responsibility, a compassionate economy, and a just society in the San Francisco
Bay Area, across the nation, and around the world. Projects
include monitoring the cleanup of Hunters Point Shipyard,
restoring the Yosemite Slough on San Francisco Bay, and
working for the cleanup and sustainable community development of other military sites on and around the bay's
waters. This is a community-based grassroots organization.

Baja California Coastkeeper
www.wildcoast.net
925 Seacoast Dr.
Imperial Beach, CA 91932
ph: 619-423-8665
fax: 619-423-8488
info@wildcoast.net
Seeks to preserve endangered marine species and threatened coastal wild lands of the Californias. This group is an
international conservation team.

Batiquitos Lagoon Foundation
www.batiquitosfoundation.org
PO Box 130491
Carlsbad, CA 92013
ph: 760-931-0800
info@batiquitosfoundation.org
Works to preserve, protect, and enhance Batiquitos Lagoon,
one of the few remaining tidal wetlands on the southern
California coast.

Bay Institute of San Francisco
www.bay.org
500 Palm Dr., Ste. 200
Novato, CA 94949
ph: 415-506-0150
fax: 415-506-0155
bayinfo@bay.org
Protects and restores the ecosystems of San Francisco Bay
and the Sacramento–San Joaquin Delta, as well as the
rivers and streams that feed into the estuary.

**BAY NET–Monterey Bay National Marine Sanctuary
Volunteer Network**
www.mbay.net/~baynet
Bay Net
PO Box 51595
Pacific Grove, CA 93950
ph: 831-373-6396
milos@mbay.net
Seeks to enhance public awareness and understanding of
the Monterey Bay National Marine Sanctuary, its wildlife,
policies, and programs.

Birch Aquarium at Scripps
www.aquarium.ucsd.edu
9500 Gilman Dr., Dept 0207
La Jolla, CA 92093-0207
ph: 858-534-FISH
fax: 858-534-7114
aquariuminfo@ucsd.edu
Provides ocean science education and promotes conservation. The aquarium is the interpretive center for the Scripps Institution of Oceanography in La Jolla, California.

Bluevoice.org
www.bluevoice.org
1252 B St.
Petaluma, CA 94952
contact@bluevoice.org
Produces films and news reports aimed at protecting oceans, whales, and dolphins by documenting tragic events in the marine world.

Bluewater Network
www.bluewaternetwork.org
311 California St., Ste. 510
San Francisco, CA 94104
ph: 415-544-0790
fax: 415-544-0796
bluewater@bluewaternetwork.org
Champions innovative solutions and inspires individuals to protect earth's finite and vulnerable ecosystems. Bluewater has been a leader in identifying and reforming marine engine and ship-based sources of pollution, including air pollution, water pollution, and invasive species.

Bolinas Lagoon Foundation
www.bolinaslagoon.org
PO Box 444
Stinson Beach, CA 94970
ph: 415-868-0518
m.mace@worldnet.att.net
Promotes conservation and protection of Bolinas Lagoon and its natural surroundings in Marin County, California, as a unique national resource. The foundation also supports scientific study of the lagoon, encourages an active concern in the public, and works with and assists other nonprofit and charitable agencies concerned with the welfare of Bolinas Lagoon.

Bolsa Chica Land Trust
www.bolsachicalandtrust.org
5200 Warner Ave., #108
Huntington Beach, CA 92649
ph: 714-846-1001
bclt@bolsachicalandtrust.org

Leads the effort to purchase and preserve the remaining open space in Bolsa Chica, one of the last wetland ecosystems in Southern California.

Cabrillo Marine Aquarium
www.cabrilloaq.org
3720 Stephen White Dr.
San Pedro, CA 90731
ph: 310-548-7562
fax: 310-548-2649
info@cabrilloaq.org
Promotes knowledge and awareness of marine life in Southern California through recreational, educational, and research programs.

California Academy of Sciences
www.calacademy.org
875 Howard St.
San Francisco, CA 94103-3009
ph: 415-321-8000
info@calacademy.org
Explores, explains, and protects the natural world and educates the public about California's ocean resources through the Steinhart Aquarium, sustainable seafood guide, and other programs. Founded in 1853, the academy was the first scientific institution in the western United States.

California Coastkeeper Alliance
www.cacoastkeeper.org
PO Box 3156
Fremont, CA 94539
ph: 510-770-9764
lsheehan@cacoastkeeper.org
Seeks to protect and restore the quality of California's aquatic ecosystem. A coalition of local waterkeepers, the alliance ensures that California's coastkeepers are able to coordinate their varied efforts.

California League of Conservation Voters
www.ecovote.org
1212 Broadway, Ste. 630
Oakland, CA 94612
ph: 510-271-0900
fax: 510-271-0901
ecovote@ecovote.org
Protects the environmental quality of the state by increasing awareness of the environmental performance of all elected officials. The league works to elect environmentally responsible candidates and holds them accountable to the environmental agenda once elected.

California Wilderness Coalition
www.calwild.org
1212 Broadway, Ste. 1700
Oakland, CA 94612
ph: 510-451-1450
fax: 510-451-1445
info@calwild.org
Works for passage of legislation designating new wilderness areas, including coastal areas and wild and scenic rivers.

Californians Organized to Acquire Access to State Tidelands (COAAST)
1475 Los Alamos Rd.
Santa Rosa, CA 95409
ph: 707-539-0153
crhinehart@aol.com
Fights for public access and coastal open space easements. This group is an alliance of several coastal organizations.

Catalina Island Marine Institute
www.guideddiscoveries.org/cimisite/school.htm
C/O Guided Discoveries, Inc.
PO Box 1360
Claremont, CA 91711
ph: 800-645-1423
info@GuidedDiscoveries.org
Provides teachers with an opportunity to let students experience, firsthand, marine biology and island ecology by immersing them in experiential activities with an educational focus. CIMI serves over 400 school groups at three different sites on Catalina Island, and also offers an Expeditionary Tall Ship program.

Center for Ecosystem Survival
www.savenature.org
699 Mississippi St., Ste. 106
San Francisco, CA 94107
ph: 415-648-3392
fax: 415-648-3392
info@savenature.org
Creates global partnerships for biological diversity through worldwide ecosystem protection programs such as "Adopt a Reef." The center's goals are to increase individual participation in and awareness of worldwide conservation by providing an opportunity for direct action.

Center for Marine Biodiversity and Conservation
www.cmbc.ucsd.edu
Scripps Institution of Oceanography
University of California–San Diego
La Jolla, CA 92093
ph: 858-822-2790
fax: 858-822-1267
cmbc@ucsd.edu
Seeks to increase public understanding of scientific issues relating to the protection and restoration of marine biodiversity.

Central California Council of Diving Clubs Inc.
www.cencal.org
4016 Walnut Blvd.
Walnut Creek, CA 94596
ph: 925-932-8838
scampi@campin.net
Promotes the principles of wise and equitable ocean legislation, safety, conservation, access, sportsmanship, and underwater sports and furthering knowledge of marine phenomena.

Central Coast Salmon Enhancement
www.centralcoastsalmon.com
PO Box 277
Avila Beach, CA 93424
ph: 805-473-8221
fax: 805-473-8167
salmonfix@aol.com
Promotes enhancement and restoration of the Central Coast salmon fishery and local creeks.

Clean Water Action–San Francisco
www.cleanwateraction.org/ca/index.htm
111 Montgomery St., Ste. 600
San Francisco, CA 94105
ph: 415-369-9160
fax: 415-369-9180
cwasf@cleanwater.org
Works for clean, safe, and affordable water, water pollution prevention, creation of environmentally safe jobs and businesses, and empowerment of people to make democracy work. This group is part of a national citizen's organization.

Coast Action Group
www.mecgrassroots.org/groups/CAG.html
Point Arena, CA 95468
ph: 707-882-2484
alevine@mail.mcn.org
Works to preserve the riverine and coho salmon runs on the Garcia River in southern Mendocino County.

Coastal Conservancy
www.coastalconservancy.ca.gov
1330 Broadway, 11th Fl.
Oakland, CA 94612
ph: 510-286-1015
fax: 510-286-0470
jbalanda@scc.ca.gov
Works to preserve, protect, and restore the resources of the
California coast. The vision of this state agency is for a
beautiful, restored, and accessible coastline.

Coastal Watershed Council
www.coastal-watershed.org
PO Box 1459
Santa Cruz, CA 95061
ph: 831-464-9200
fax: 831-464-9214
tcdoan@coastalws.org
Promotes coastal watershed protection through monitoring,
research, education, and stewardship.

Coastkeeper–Orange County
www.coastkeeper.org
441 Newport Blvd., Ste. 103
Newport, CA 92663
ph: 949-723-5424
fax: 949-675-7091
coastkeeper1@earthlink.net
Protects and preserves Orange County's marine habitat and
watersheds through education, advocacy, restoration, and
enforcement.

Coastside Habitat Coalition
www.isant.com/chc_home.htm
PO Box 49
San Gregorio, CA 94074
ph: 650-726-2499
abisart@adelphia.net
Provides stream monitoring and education for preservation
and protection of coastal habitats in the San Gregorio
region.

Coastwalk
www.coastwalk.org
825 Gravenstein Hwy. North #8
Sabastopol, CA 95472
ph: 800-550-6854
fax: 707-829-0326
coastwalk@coastwalk.org
Helps people to experience the California coast in an inti-
mate and respectful way. Coastwalk envisions a completed
California Coastal Trail from Oregon to the Mexican border
while preserving the fragile coastal environment. It also
coordinates government, community, and trail organizations
to promote the California Coastal Trail, protect access to
the trail, and encourage coastal protection.

Community Environmental Council
www.communityenvironmentalcouncil.org/wrc
www.watershedresourcecenter.org
Southcoast Watershed Resource Center
2981 Cliff Dr.
Santa Barbara, CA 93109
ph: 805-682-6113
fax: 805-682-8113
mmowers@cecmail.org
Focuses on the connection between healthy watersheds
and personal habits. The center promotes actions that individuals can take such as cleaning up after pets, landscaping
with native plants, and properly disposing of everyday
chemicals.

Conception Coast Project
www.conceptioncoast.org
PO Box 80241
Goleta, CA 93118
ph: 805-637-1200
studarus@conceptioncoast.org
Works to protect and restore the natural heritage of the
Conception Coast region (central California) through science planning and community involvement.

Coral Reef Alliance
www.coral.org
417 Montgomery St., Ste. 205
San Francisco, CA 94104
ph: 415-834-0900
fax: 415-834-0999
info@coral.org
Promotes coral reef conservation around the world by
working with the diving industry, government, local communities, and other organizations to protect and manage
coral reefs, establish marine parks, fund conservation
efforts, and raise public awareness with the mission to keep
coral reefs alive for future generations.

Defend the Bay
www.defendthebay.org
PO Box 7609
4101 Birch St., Ste. 150
Newport Beach, CA 92658
ph: 949-722-7822
fax: 949-722-6911
info@defendthebay.org
Works to protect Newport Bay and the coastal waters in
Orange County. This organization employs a unique mix of
education, science, and law as tools to preserve the county's heritage and environmental crown jewels.

Defenders of Wildlife
www.defenders.org/wildlife/new/marine.html
Marine Program
PO Box 959
Moss Landing, CA 95039
ph: 831-726-9010
fax: 831-726-9020
jcurland@defenders.org
Works to protect all native wild animals and plants in their
natural communities, including marine plants and animals
in coastal and offshore habitats.

Dive Equipment Manufacturers Association
www.dema.org
3750 Convoy St., Ste. 310
San Diego, CA 92111-3741
ph: 858-616-6408
fax: 858-616-6495
info@dema.org
Promotes sustainable growth in safe recreational diving
and snorkeling while protecting the underwater
environment.

Dunes Center
www.dunescenter.org
1055 Guadalupe St.
Guadalupe, CA 93434
ph: 805-343-2455
fax: 805-343-0442
info@dunescenter.org
Conducts research and restoration projects related to
Guadalupe and Nipomo Dunes.

Earth Island Institute
www.earthisland.org
300 Broadway, Ste. 28
San Francisco, CA 94133-3312
ph: 415-788-3666
fax: 415-788-7324
www.earthisland.org/feedback.html
Works for solutions to environmental problems by promoting citizen action and incubating a diverse network of projects, many of which are marine oriented.

Earthjustice Legal Defense Fund–California
www.earthjustice.org/regional/oakland
426 17th St., 5th Fl.
Oakland, CA 94612-2820
ph: 510-550-6725
fax: 510-550-6749
eajusca@earthjustice.org
Works to protect ecosystems and public lands and waters
and to promote healthy communities in California by providing free legal services to national organizations such as
the Sierra Club and regional groups such as the San
Francisco Baykeeper.

Environment California–Los Angeles
www.environmentcalifornia.org
3435 Wilshire Blvd., Ste. 385
Los Angeles, CA 90010
ph: 213-251-3688
info@environmentcalifornia.org
Works to give the people of California a voice on a wide
range of critically important environmental issues such as
coastal protection, especially when powerful special interests stand in the way of reform.

Environment California–Sacramento
www.environmentcalifornia.org
Legislative Office
1107 9th St., Ste. 601
Sacramento, CA 95814
ph: 916-446-8062
info@environmentcalifornia.org

Environment California–San Francisco
www.environmentcalifornia.org
369 Broadway, Ste. 200
San Francisco, CA 94133
ph: 415-622-0086
info@environmentcalifornia.org

Environment California–Santa Barbara
www.environmentcalifornia.org
1129 State St., #10
Santa Barbara, CA 93101
ph: 805-568-3832
info@environmentcalifornia.org

Environment Now
www.Environmentnow.org
2515 Wilshire Blvd.
Santa Monica, CA 90403
ph: 310-829-5568
fax: 310-829-6820
jgarvender@environmentnow.org
Leads the creation of measurably effective environmental
programs to protect and restore California's environment by
focusing on preservation of coasts and forests and reduction of air pollution and urban sprawl.

Environmental Center of San Luis Obispo County (ECOSLO)
www.ecoslo.org
1204 Nipomo St.
San Luis Obispo, CA 93401
ph: 805-544-1777
fax: 805-544-1871
info@ecoslo.org
Works to protect and enhance the natural environment and human well-being through education, advocacy, and direct action. The center serves as a catalyst for community environmental stewardship, nurturing a sustainable human habitat in harmony with natural systems, now and for future generations.

Environmental Defense
www.environmentaldefense.org
5655 College Ave.
Oakland, CA 94618
ph: 510-658-8008
fax: 510-658-0630
members@environmentaldefense.org
Fights to protect human health, restore the oceans and ecosystems, and curb global warming. California focuses include establishing a network of marine-protected areas.

Environmental Forum of Marin
www.marinefm.org
PO Box 150459
San Rafael, CA 94915
ph: 415-479-7814
forum@marinefm.org
Seeks to protect and enhance the environment by educating its members and the Marin citizenry about environmental issues.

Environmental Health Coalition
www.environmentalhealth.org
401 Miles of Carsway, Ste. 310
National City, CA 91950
ph: 619-474-0220
fax: 619-474-1210
ehc@environmentalhealth.org
Works to achieve environmental and social justice by empowering communities to act together to make social change. The coalition fights toxic pollution, protects public health, and promotes environmental justice. Areas of concern include the U.S. Navy's dredging of toxic and hazardous sediments in the harbor and the selection of San Diego for the homeporting of additional nuclear aircraft carriers.

Environmental Volunteers
www.Evols.org
3921 E. Bayshore Rd.
Palo Alto, CA 94303-4326
ph: 650-961-0545
fax: 650-961-0548
info@evols.org
Promotes the understanding of and responsibility for the environment through hands-on science education on and around bay and ocean waters.

Farallones Marine Sanctuary Association
www.farallones.org
The Presidio
PO Box 29386
San Francisco, CA 94129
ph: 415-561-6625
fax: 415-561-6616
Strives to protect sanctuary wildlife through development of a diverse community of informed and active ocean sanctuary stewards.

Friends of Moss Landing Marine Labs
http://friends.mlml.calstate.edu
8272 Moss Landing Rd.
Moss Landing, CA 95039
ph: 831-771-4100
fax: 831-632-4403
friends@mlml.calstate.edu
Serves as a liaison between MLML and the community at large to support the research, education, and conservation work of the laboratories, fostering such support through education programs, events, fund-raising activities, and an active alumni association.

Friends of the Dunes
www.friendsofthedunes.org
PO Box 186
Arcata, CA 95518
ph: 707-444-1397
fax: 707-444-0447
info@friendsofthedunes.org
Seeks to conserve the natural diversity of coastal, bay, and dune environments through community-supported education and stewardship programs.

Friends of the Estuary at Morro Bay
www.friendsofestuary.org
PO Box 1375
Morro Bay, CA 93443-1375
Seeks to restore, maintain, and enhance the estuary at Morro Bay. Friends is a nonprofit organization dedicated to this national treasure.

Friends of the Los Angeles River
www.folar.org
570 W. Ave. 26 #250
Los Angeles, CA 90065
ph: 323-223-0585
fax: 323-223-2289
mail@folar.org
Seeks to revitalize and protect the Los Angeles River, one of
the most abused and degraded urban rivers in the world,
and to create a Los Angeles River greenway from the
mountains to the sea. FoLAR is a nonprofit grassroots citizens group founded in 1986.

Friends of Newport Coast
PO Box 671
Corona Del Mar, CA 92625
ph: 949-644-5998
fax: 949-759-1067
fernpirk@yahoo.com
Seeks to preserve and protect the maximum area of open
space and wildlife habitat possible on the Newport coast
for public use and enjoyment.

Friends of the River
www.friendsoftheriver.org
915 20th St.
Sacramento, CA 95814
ph: 916-442-3155
fax: 916-442-3396
info@friendsoftheriver.org
Works to preserve, protect, and restore California's rivers,
streams, and watersheds.

Friends of the Sea Otter
www.seaotters.org
125 Ocean View Blvd., Ste. 204
Pacific Grove, CA 93950
ph: 831-373-2747
fax: 831-373-2749
info@seaotters.org
Works actively with state and federal agencies to maintain
the current protections for sea otters as well as to increase
and broaden these preservation efforts.

Gaviota Coast Conservancy
www.gaviotacoastconservancy.org
PO Box 1099
Goleta, CA 93116
ph: 805-683-6631
gavcoast@silcom.com
Provides permanent protection of Gaviota Coast's unique
natural, scenic, agricultural, recreational, and cultural
resources.

Golden Gate Fishermen's Association
www.sfsportfishing.com/GGFA
50 Briarwood
San Rafael, CA 94901
Preserves and enhances California's fisheries and natural
resources.

Greenpeace–California
www.greenpeaceusa.org
75 Arkansas St., Ste. 1
San Francisco, CA 94107-2434
ph: 415-255-9221
fax: 415-255-9201
Uses nonviolent, creative confrontation to expose global
environmental problems and force solutions for a green
and peaceful future. Greenpeace's goal is to ensure the
earth's ability to nurture life in all its diversity.

Habitat Media
www.habitatmedia.org
734 A St.
San Rafael, CA 94901
ph: 415-458-1696
fax: 415-458-1695
info@habitatmedia.org
Specializes in producing engaging media that educates
viewers about the need for sustainable development and
inspires participation with a particular focus on marine
issues including two PBS documentaries: *Empty Oceans,
Empty Nets* and *Farming the Seas*.

Headlands Institute
www.yni.org/hi
Golden Gate National Rec. Area, Bldg. 1033
Sausalito, CA 94965
ph: 415-332-5771
fax: 415-332-5784
hi@yni.org
Provides educational adventures in nature's classroom to
inspire a personal connection to the natural world and
responsible actions to sustain it. The institute is located on
the beach.

Heal the Bay
www.healthebay.org
3220 Nebraska Ave.
Santa Monica, CA 90404
ph: 310-453-0395
fax: 310-453-7927
info@healthebay.org
Uses research, education, community action, and policy
programs to ensure that Santa Monica Bay and Southern
California coastal waters are safe and healthy for people
and marine life.

Heal the Ocean
www.healtheocean.org
PO Box 90106
Santa Barbara, CA 93190
ph: 805-965-7570
fax: 805-962-0651
info@healtheocean.org
Deals swiftly with the critical issue of ocean pollution that
closes Santa Monica beaches. This independent organization uses its own tests to evaluate the results of county,
state, and federal action. It works directly with property
owners for solutions to problems and assesses science and
legal action on behalf of ocean protection.

Humboldt Baykeeper
424 First St.
Eureka, CA 95501
ph: 707-268-8900 x2
ecorights@earthlink.net
Seeks to protect and enhance the quality of aquatic
resources for the bay.

Institute for Fisheries Resources
www.ifrfish.org
PO Box 29196
San Francisco, CA 94129-0196
ph: 415-561-3474
fax: 415-561-5464
zgrader@ifrfish.org
Works to protect and restore fish resources and the human
economies that depend on them. IFR unites resource stakeholders, protects fish populations, and restores aquatic
habitats by establishing alliances among fishermen and
women, government agencies, and concerned citizens.

International Longshore and Warehouse Union
www.ilwu.org
1188 Franklin St., 4th Fl.
San Francisco, CA 94109
ph: 415-775-0533
fax: 415-775-1302
info@ilwu.org
Works with other groups and individuals involved in protecting maritime communities and resources. The ILWU has
a long history representing workers in the ports. Its 42,000
members span the West Coast, Alaska, and Hawaii and are
united in one organization regardless of religion, race,
creed, color, sex, political affiliation, or nationality.

International Marine Mammal Project
www.earthisland.org/immp
300 Broadway, Ste. 28
San Francisco, CA 94113
ph: 415-788-3666
fax: 415-788-7324
marinemammal@earthisland.org
Works to make oceans safe for marine mammals worldwide. The IMMP strives to eliminate dolphin mortality
caused by the international tuna fishing industry, to end the
use of drift nets, and to stop the purse-seine fishers from
encircling dolphins in their nets. In addition, the project
aims to stop the resumption of commercial whaling worldwide, to promote sustainable fishing, and to protect the
habitat of whales, dolphins, and other marine species.

International Rivers Network
www.irn.org
1847 Berkeley Way
Berkeley, CA 94703
ph: 510-848-1155
fax: 510 848 1008
info@irn.org
Supports local communities working to protect their rivers
and watersheds. The organization works to halt destructive
river development projects and encourages equitable and
sustainable methods of meeting needs for water, energy,
and flood management.

International Surfing Museum
www.surfingmuseum.org
411 Olive Ave.
Huntington Beach, CA 92648
ph: 714-960-3483
intsurfing@earthlink.net
Features some of the most significant artifacts in the history
of surfing and references efforts to protect standing waves
from marina development and other threats.

Intersea Foundation
www.intersea.org
PO Box 1106
Carmel Valley, CA 93924
ph: 831-659-5807
fax: 831-659-5807
info@intersea.org
Conducts whale research expeditions in Southeast Alaska
and other parts of the world. These high-adventure natural
history and research voyages enable laypersons to assist
scientists in the study of cetaceans and their environment.

La Jolla Friends of the Seals
www.lajollaseals.com
PO Box 2016
La Jolla, CA 92038
ph: 619-687-3588
Seeks to protect and preserve the La Jolla Harbor seal
colony for ecological, educational, scientific, historic, and
scenic opportunities through a naturalist-docent program
that educates the public and gathers research data on the
seals. Friends was established in 1999.

League of Women Voters
www.lwvc.org
801 12th St., Ste. 220
Sacramento, CA 95814
ph: 916-442-7215
fax: 916-442-7362
lwvc@lwvc.org
Encourages informed and active participation of citizens in
government. Formed in 1920, the nonpartisan organization
went to court in 2004 to assure a fair vote count for ocean
activist Donna Frye in her write-in campaign for mayor of
San Diego.

Long Beach Department of Public Works
City Hall, 9th Fl.
333 West Ocean Blvd.
Long Beach, CA 90802
ph: 562-570-6023
fax: 562-570-6023
Seeks innovative ways to reduce stormwater runoff. The
department is deemed a model municipal government
agency in this effort.

Long Beach Marine Institute
www.longbeachmarineinst.com
Box 281
6475 E. Pacific Coast Hwy.
Long Beach, CA 90803
ph: 562-431-7156
info@longbeachmarineinst.com

Works to better acquaint local residents with the valuable
resources of the Southern California coast. The association
of marine biology and oceanography researchers and educators seeks to help people understand the process of
marine science and hopes to spark an interest in future
exploration of the ocean realm.

The Manta Network
www.Save-the-Mantas.org
326 Pacheco Ave.
Santa Cruz, CA 95062
ph: 831-426-4400
help@Save-the-Mantas.org
Works for the protection and conservation of manta and
mobula rays worldwide. Educates the fishing and tourism
industries, government organizations, conservation groups,
and the public about the importance of this species. Works
with volunteers and researchers in more than 15 countries
to establish their importance within the ocean's ecosystem.

Marine Mammal Center
www.tmmc.org
1065 Fort Cronkhite
Sausalito, CA 94965-2609
ph: 415-289-7325
fax: 415-289-7333
admin@tmmc.org
Works to conserve habitat and on the rescue and humane
treatment of ill, injured, and orphaned marine mammals
and to return healthy animals to the wild. The center recognizes human interdependence with marine mammals and
their importance as sentinels of the ocean environment, the
health of which is essential for all life.

Marine Mammal Care Center at Fort MacArthur
3601 S. Gaffney St.
San Pedro, CA 90731
ph: 310-548-5677
fax: 310-548-6394
FortMacCLs@aol.com
The Center supports marine animal rehabilitation along the
Southern California coast.

Marine Science Institute
www.sfbaymsi.org
500 Discovery Parkway
Redwood City, CA 94063
ph: 650-364-2760
gail@sfbaymsi.org
Offers hands-on science and environmental education programs to students of all ages throughout the Bay Area and
northern California.

Mendocino Environmental Center

www.mecgrassroots.org/mecframe2.html
106 West Standley
Ukiah, CA 95482
ph: 707-468-1660
fax: 707-462-2370
Works through educational outreach, nonviolent direct action, and the legal system to uphold and promote environmental and social justice in coastal Mendocino County and beyond.

Monterey Academy of Oceanographic Science

www.maos.montereyhigh.com/
101 Herrmann Dr.
Monterey, CA 93940
ph: 831-392-3858
fax: 831-649-3841
maos@mpusd.k12.ca.us
Brings motivation to public school education by emphasizing four years of math, science, and English broadly, but not exclusively, applied to scientific and oceanographic research. The academy contributes to the local community sense of the Monterey Bay as a global center for marine science, conservation, and exploration.

Monterey Bay Aquarium

www.mbayaq.org
886 Cannery Row
Monterey, CA 93940
ph: 831-648-4800
fax: 831-648-4810
Seeks to inspire conservation of the oceans. Located in an old Sardine cannery, the aquarium is world famous for its scenic locale and innovative exhibits.

Monterey Bay Aquarium Research Institute

www.mbari.org
7700 Sandholdt Rd.
Moss Landing, CA 95039-9644
ph: 831-775-1700
fax: 831-775-1620
info@mbari.org
Achieves and maintains a position as a world center for advanced research and education in ocean science and technology through the development of better instruments, systems, and methods for scientific research in the deep waters of the ocean.

Monterey Bay Sanctuary Foundation

www.mbnmsf.org
299 Foam St.
Monterey, CA 93940
ph: 831-647-4209
fax: 831-647-4244
admin@mbnmsf.org
Advances understanding of the Monterey Bay National Marine Sanctuary (see Section 4).

Moss Landing Marine Laboratories

www.mlml.calstate.edu
8272 Moss Landing Rd.
Moss Landing, CA 95039
ph: 831-771-4400
fax: 831-632-4403
frontdesk@mlml.calstate.edu
Equips the pioneers of the future through a hands-on, field-oriented approach to a curriculum that places students at the frontiers of where marine science discoveries are being made.

National OCS Coalition

6947 Cliff Ave.
Bodega Bay, CA 94923
ph: 707-875-2345
waterway@moniter.net
Links activists, the public, and elected officials to prevent reckless offshore oil and gas development and to maintain watch over plans for ocean mining on the Outer Continental Shelf (OCS).

Natural Heritage Institute

www.n-h-i.org
100 Pine St., Ste. 1550
San Francisco, CA 94111
ph: 415-693-3000
fax: 415-693-3178
nhi@n-h-i.org
Restores natural systems, including marine systems, for an independent world. A group of experienced conservation lawyers and scientists who saw a need for a toolkit for the next era of environmental problem solving founded this nonprofit organization in 1989.

Natural Resources Defense Council–San Francisco

www.nrdc.org
111 Sutter St., 20th Fl.
San Francisco, CA 94104
ph: 415-875-6100
fax: 415-875-6161
nrdcinfo@nrdc.org
Uses law, science, and the support of more than one million online activists to protect the planet's wildlife and wild places, and to ensure a safe and healthy environment for all living things. NRDC also has an active marine program.

Natural Resources Defense Council–Los Angeles

www.nrdc.org
1314 Second St.
Santa Monica, CA 90401
ph: 310-434-2300
fax: 310-434-2399
nrdcinfo@nrdc.org

Newport Bay Naturalists and Friends

www.newportbay.org
600 Shellmaker Rd.
Newport Beach, CA 92660
ph: 949-640-6746
fax: 949-642-3189
info@newportbay.org
Restores and preserves the native habitat of the bay and its surroundings; educates the public about the ecological value of the bay; and tries to achieve good water quality, healthy native flora and fauna, and compatible public use of a protected ecosystem in a dense urban environment.

Newport Harbor Nautical Museum

www.nhnm.org
151 East Pacific Coast Highway
Newport Beach, CA 92660
ph: 949-673-3377
info@nhnm.org
Preserves the maritime history of Newport Harbor and Southern California.

Northcoast Marine Mammal Center

www.northcoastmarinemammal.org
424 Howe Dr.
Crescent City, CA 95531
ph: 707-465-6265
Rescue@northcoastmarinemammal.org
Rescues and rehabilitates stranded seals, sea lions, dolphins, porpoises, and whales along the northernmost coast of California. The center also promotes public understanding of marine mammals and the importance of the marine environment.

O'Neill Sea Odyssey

www.oneillseaodyssey.org
2222 East Cliff Dr., Ste. 222
Santa Cruz, CA 95062
ph: 831-465-9390
fax: 831-462-9188
dhaifley@oneillseaodyssey.org
Provides a hands-on educational experience to encourage the protection and preservation of the living sea and community.

Ocean Champions

www.oceanchampions.org
202 San Jose Ave.
Capitola, CA 95010
ph: 831-462-2539
fax: 831-462-2542
dave@oceanchampions.org
Works to support proven leaders and elect new candidates to the U.S. Congress who will advocate for ocean conservation. Ocean Champions is a nonpartisan organization.

The Ocean Conservancy–Pacific Region Office

www.oceanconservancy.org/dynamic/aboutUs/offices/
pacific/pacific.htm
116 New Montgomery St., Ste. 10
San Francisco, CA 94105
ph: 415-979-0900
fax: 415-979-0901
wchabot@oceanconservancy.org
Uses science-based advocacy, research, and public education to inform, inspire, and empower people to speak and act for the oceans. The Ocean Conservancy is a national organization dedicated to protecting ocean ecosystems. The organization has a national office in Washington, D.C., and ten regional offices, including three in California. Its major programs include coral reef protection, debris monitoring, stormwater runoff and pollution, and coastal cleanup.

The Ocean Conservancy–Santa Barbara Field Office

www.oceanconservancy.org
714 Bond Ave.
Santa Barbara, CA 93101
ph: 805-687-2322
fax: 805-687-5635
ghelms@psinet.com

The Ocean Conservancy–Santa Cruz

www.oceanconservancy.org
55 C Municipal Wharf
Santa Cruz, CA 95060
ph: 831-425-1363
fax: 831-425-5604
kgaffney@psinet.com

Ocean Conservation Society

www.oceanconservation.org
PO Box 12860
Marina del Rey, CA 90295
ph: 310-822-5205
fax: 310-822-5729
info@oceanconservation.org
Works to increase public awareness of the problems facing
marine ecosystems through the development and dissemination of educational programs such as the Los Angeles
Dolphin Project and materials based on research the society
conducts. OCS provides volunteer field programs that
include academic internships.

Ocean Futures Society

www.oceanfutures.com
325 Chapala St.
Santa Barbara, CA 93101
ph: 805-899-8899
fax: 805-899-8898
contact@oceanfutures.org
Seeks to explore the global ocean, inspiring and educating
people throughout the world to act responsibly for its protection, documenting the critical connection between
humanity and nature, and celebrating the ocean's vital
importance to the survival of all life on this planet. Jean-
Michel Cousteau is the society's founder.

Ocean Institute

www.ocean-institute.org
24200 Dana Pt. Harbour Dr.
Dana Point, CA 92629
ph: 949-496-2274
fax: 949-248-5557
oi@ocean-institute.org
Works to inspire all generations through education to
become responsible stewards of the world's oceans.

Ocean Protection Coalition

PO Box 1265
Mendocino, CA 95460
ph: 707-937-2050
Promotes protecting oceans and coastal and land areas,
sharing information on what is needed to do so from a
local, grassroots perspective.

Oceana–Los Angeles

www.oceana.org
501 Santa Monica Blvd., Ste. 312
Santa Monica, CA 90401
ph: 310-899-3026
fax: 310-899-3027
socal@oceana.org
Campaigns to protect and restore the world's oceans by
using teams of marine scientists, economists, lawyers, and
advocates to help enact specific and concrete policy
changes to reduce pollution and to prevent the collapse of
fish populations, marine mammals, and other sea life.

Otter Project

www.otterproject.org
3098 Stewart Ct.
Marina, CA 93933
ph: 831-883-4159
exec@otterproject.org
Promotes the rapid recovery of the California sea otter, an
indicator of nearshore ocean health, by facilitating research
and communicating research results to the general public
and policy makers.

Pacific Cetacean Group

www.pacificcetaceangroup.org
PO Box 835
Moss Landing, CA 93933
info@pacificcetaceangroup.org
Strives to increase the knowledge of marine mammals and
promote conservation of their habitats through scientific
research and education.

Pacific Coast Federation of Fishermen's Associations (PCFFA)–Southwest Office/Institute for Fisheries Resources

www.pcffa.org
PO Box 29370
San Francisco, CA 94129-0370
ph: 415-561-5080
fax: 415-561-5464
fish1ifr@aol.com
Leads the commercial fishing industry in ensuring the rights
of individual fishermen and fighting for the long-term survival of commercial fishing as a productive livelihood and
way of life. With 1,800 members, the PCFFA is the largest
and most politically active trade association of commercial
fishermen on the West Coast.

Pacific Environment
www.pacificenvironment.org
311 California St., Ste. 650
San Francisco, CA 94104-2608
ph: 415-399-8850
fax: 415-399-8860
info@pacificenvironment.org
Protects the living environment of the Pacific Rim by
strengthening democracy, supporting grassroots activism,
empowering communities, and redefining international policies. For nearly two decades this group has been at the
forefront of efforts to protect and preserve the ecological
treasures of the Pacific Rim.

**Pacific Marine Conservation Council–California
Office**
www.pmcc.org
PO Box 327
Arcata, CA 95518
ph: 707-822-4667
fax: 707-822-4667
Promotes initiatives in the community-based management
arena, and advocates cooperative fisheries research and
science-based policy. PMCC's Rockfish Campaign is
designed to rebuild West Coast rockfish stocks and advocate stronger socioeconomic support for fishing families.

Pacific Ocean Conservation Network
580 Market St., Ste. 550
San Francisco, CA 94104
Seeks to protect ocean resources and habitat sustainable
for future generations. The network represents a coalition
of conservation, science, and fishing industry leaders.

Pacific Seabird Group
www.pacificseabirdgroup.org
c/o Ron LeValley
920 Samoa Blvd., Ste. 210
Arcata, CA 95521
info@pacificseabirdgroup.org
A society of professional seabird researchers and managers
dedicated to the study and conservation of seabirds.

Pacific Wildlife Project
www.PACIFICWILDLIFE.org
PO Box 7673
Laguna Niguel, CA 92607-7673
ph: 949-831-1178
fax: 949-249-2709
info@ pacificwildlife.org
Works to protect and preserve wildlife through education
community outreach and rescue and medical treatment of
all wild species.

Pacifica Ocean Discovery Center
www.oceandiscoverycenter.org
PO Box 914
Pacifica, CA 94044
ph: 650-922-2043
aaron@oceandiscoverycenter.org
Works to preserve the ocean and coastal environment of
San Mateo County through education, research, and conservation.

PADI–Professional Association of Dive Instructors
www.padi.com
30151 Tomas St.
Rancho Santa Margarita, CA 92688-2125
ph: 949-858-7234
fax: 949-858-7264
webmaster@padi.com
Works on educational development of scuba diving professionals and enthusiasts and supports various marine conservation efforts through its educational foundation.

Pelagic Shark Research Foundation
www.pelagic.org, www.pelagic.org/archive
750 Bay Ave., Unit 216
Capitola, CA 95010
ph: 831-459-9346
psrf@pelagic.org
Promotes education about and conservation of sharks.

Petaluma Riverkeeper
521 Walnut St.
Petaluma, CA 94952
ph: 707-763-2310
petaluma@baykeeper.org
Strives to protect and enhance the quality of aquatic
resources on the Petaluma River.

Planning and Conservation League
www.pcl.org
921 11th St., Ste. 300
Sacramento, CA 95814
ph: 916-444-8726
fax: 916-448-1789
pclmail@PCL.org
Lobbies the California State Legislature on a full range of
environmental issues and sponsors environmental initiatives. An official resolution of the legislature stated that
"participation on every key environmental issue before the
State Legislature has demonstrated PCL's effectiveness in
preserving the quality of life for all Californians."

Point Reyes Bird Observatory
www.prbo.org
4990 Shoreline Highway
Stinson Beach, CA 94970
ph: 415-868-1221
fax: 415-868-1946
prbo@prbo.org
Works to conserve birds, other wildlife, and ecosystems
through innovative scientific research and outreach. PRBO
biologists study birds to protect and enhance biodiversity in
marine, terrestrial, and wetland systems in western North
America.

Project AWARE
www.projectaware.org
30151 Tomas St. Ste. 200
Rancho Santa Margarita, CA 92688-2125
ph: 1-866-80-AWARE; 949-858-7657 x2659
information@projectaware.org
Works on aquatic conservation on a global scale and provides support for ongoing education and awareness programs, environmental activities, and habitat research.

Public Media Center
www.publicmediacenter.org
466 Green St.
San Francisco, CA 94133
ph: 415-434-1403
fax: 415-986-6779
info@publicmediacenter.org
Produces print and broadcast campaigns for social change
on behalf of a large number of citizen activists including
coastal and ocean groups.

Reef Check Foundation
www.reefcheck.org
PO Box 1057
17575 Pacific Coast Highway
Pacific Palisades, CA 90272-1057
ph: 310-230-2371
fax: 310-230-2376

rcinfo@reefcheck.org
Works with communities, governments, and businesses to
educate the public about the coral reef crisis; to create a
global network of volunteer teams trained in Reef Check's
scientific methods who regularly monitor and report on reef
health; to facilitate collaboration that produces ecologically
sound and economically sustainable solutions; and to stimulate local community action to protect remaining pristine
reefs and rehabilitate damaged reefs worldwide. Reef
Check is an international program.

Reef Protection International
www.reefprotect.org
300 Broadway, Ste. 28
San Francisco, CA 94133
ph: 415-699-2091
fax: 415-788-7324
info@reefprotect.org
Works for coral reef conservation by supporting reform in
the saltwater aquarium trade and works to educate and
transform consumer demand in the United States. Reef
Protection International is a nonprofit project of the Earth
Island Institute.

Roundhouse Marine Studies Lab and Aquarium
www.roundhousemb.com
PO Box 1
Manhattan Beach, CA 90267
ph: 310-379-8117
fax: 310-937-9366
Roundhouse.aquarium@verizon.net
Features ten large aquaria displaying local fish and invertebrates in a facility located at the end of the Manhattan
Beach Pier. Roundhouse also holds classes and field trips
for thousands of elementary, middle school, and high
school students every year. Roundhouse serves as the educational facility for the nonprofit Oceanographic Teaching
Station.

**Russian River Watershed Protection Committee
(RRWPC)**
www.envirocentersoco.org/rrwpc
PO Box 501
Guerneville, CA 95446
ph: 707-869-0410
brendarrwpc@earthlink.net
Works in the public interest on river protection issues. The
RRWPC, a nonprofit corporation, was founded in 1980 and
was originally incorporated under the name River Citizens
Sewer Committee.

Russian Riverkeeper
www.russianriverkeeper.org
PO Box 1335
Healdsburg, CA 95448
ph: 707-433-1958

fax: 707-433-1989
info@russianriverkeeper.org
Advocates compliance with river protection laws and regulations and patrols the river with a full-time Riverkeeper to document and report violations.

Sacramento–San Joaquin Deltakeeper

3536 Rainer Ave.
Stockton, CA 95204
ph: 209-464-5090
fax: 209-464-5174
Works to protect and enhance the quality of aquatic resources in the delta.

Sanctuary Cruises

www.sanctuarycruises.com
25515 Hidden Mesa Rd.
Monterey, CA 93940
ph: 831-643-0128
mail@sanctuarycruises.org
Offers year-round whale watching and special charters from the Monterey Bay National Marine Sanctuary.

San Diego Baykeeper

www.sdbaykeeper.org
2924 Emerson St., Ste. 220
San Diego, CA 92106
ph: 619-758-7743
fax: 619-758-7740
info@sdbaykeeper.org
Works to preserve the integrity of San Diego's regional water bodies by balancing outreach, education, advocacy, and litigation to increase awareness of issues and reduce water pollution.

San Diego Council of Divers

www.sddivers.com
PO Box 84778
San Diego, CA 92138-4778
info@sddivers.com
Supports education, safety, conservation, sportsmanship, the preservation of wise legislation, and the furthering of knowledge of marine phenomena.

San Diego Oceans Foundation

www.sdoceans.org
PO Box 90672
San Diego, CA 92169-2672
ph: 619-523-1903
noelie@sdoceans.org
Promotes ocean stewardship by leading community-supported projects that enhance ocean habitats and encourage sustainable use of ocean resources.

San Elijo Lagoon Conservancy

www.sanelijo.org
PO Box 230634
Encinitas, CA 92023
ph: 760-436-3944
fax: 760-944-9606
dg@sanelijo.org
Assists in preserving, protecting, and enhancing the natural resources of the San Elijo Lagoon Ecological Reserve. Formed in 1987, the Conservancy raises funds that are used to support projects that directly benefit the health of the lagoon or upgrade services available to visitors.

San Francisco Bay Conservation and Development Commission

www.bcdc.ca.gov
50 California St., Ste. 2600
San Francisco, CA 94111
ph: 415-352-3600
fax: 415-352-3606
info@bcdc.ca.gov
Works to protect the bay from planned filling by the Army Corps of Engineers. The commission, the nation's first coastal management agency, has become a model for coastal zone management efforts to protect and enhance local marine and estuarine resources.

San Francisco Baykeeper

www.sfbaykeeper.org
55 Hawthorne St., Ste. 550
San Francisco, CA 94105
ph: 415-856-0444
fax: 415-856-0443
Leo@sfbaykeeper.org
Works to enhance the water quality of the San Francisco Bay/Delta Estuary and its tributaries through patrolling, monitoring, enforcing clean water laws, watchdogging government agencies, and educating the next generation of water-quality stewards.

San Francisco Estuary Institute

www.sfei.org
7770 Pardee Ln.
Oakland, CA 94621-1424
ph: 510-746-7334
fax: 510-746-7300
Fosters development of the scientific understanding needed to protect and enhance the San Francisco Bay Estuary. The institute fills the niche between environmental science and environmental management and policy for San Francisco Estuary and its watershed.

San Francisco Maritime National Historical Park

www.nps.gov/safr
Building E Fort Mason
San Francisco, CA 94123
ph: 415-561-7000
fax: 415-556-1624
Offers educational, music, and craft programs for all ages and provides unique opportunities for docents, interns, and volunteers to learn more about the nation's maritime heritage. Located at the west end of San Francisco's Fisherman's Wharf, this park includes the fleet of national historic landmark vessels at Hyde St. Pier, a visitor center, a maritime museum, and a library/research facility.

San Francisco Ocean Film Festival

http://oceanfilmfest.org
PO Box 475668
San Francisco, CA 94147
ph: 415-310-5259
info@oceanfilmfest.org
Presents inspirational films and lively discussions on oceanography, saltwater sports, and coastal cultures. One of only two ocean film fests in the world (the other is in France), the SF Ocean Film Festival takes place every January.

San Luis Obispo (SLO) Coastkeeper

www.epicenteronline.org
EPI–Center
1013 Monterey St., Ste. 207
San Luis Obispo, CA 93401
ph: 805-781-9932
fax: 805-781-9384
Guards the environment in the public interest through protection of aquatic resources along the coast.

Santa Barbara Channelkeeper

www.sbck.org
714 Bond Ave.
Santa Barbara, CA 93103
ph: 805-563-3377
info@sbck.org
Works to protect and restore the Santa Barbara Channel and its watersheds through enforcement, citizen action, and education. The Channelkeeper is a membership- and volunteer-based nonprofit organization.

Santa Barbara Maritime Museum

www.sbmm.org
113 Harbor Way, Ste. 190
Santa Barbara, CA 93109
ph: 805-962-8404
fax: 805-962-7634
museum@sbmm.org
Offers exhibits that explore the wonders of the California coast, the beauty of its fish and wildlife, and the fascinating maritime history of its explorers and their ships.

Santa Barbara Museum of Natural History

www.sbnature.org
2559 Huesto del Sol Rd.
Santa Barbara, CA 93105
ph: 805-682-4711
fax: 805-569-3170
info@sbnature2.org
Offers access to objects and ideas related to the natural history of the Santa Barbara region, its coastline and offshore islands. Its Department of Vertebrate Zoology has long collected and analyzed data on marine mammal strandings and other marine phenomena.

Santa Catalina Island Conservancy

www.catalinaconservancy.org
PO Box 2739
Avalon, CA 90704
ph: 310-510-2595
fax: 310-510-2594
Strives to be a responsible steward of Santa Catalina Island through a balance of conservation, education, and recreation.

Santa Monica Baykeeper

www.smbaykeeper.org
PO Box 10096
Marina del Rey, CA 90295
ph: 310-305-9645
fax: 310-305-7985
info@smbaykeeper.org
Surveys the environmental health of the Santa Monica Bay, San Pedro Bay, and the adjacent coastal waters and watersheds; alerts the public to potential hazards; and exposes those who contribute in any way to the degradation of the ecosystem. The Baykeeper seeks to detect, investigate, and deter violators of environmental water quality or habitat laws.

Santa Monica Bay Restoration Commission

www.santamonicabay.org
320 West 4th St., Ste. 200
Los Angeles, CA 90013
ph: 213-576-6615
fax: 213-576-6646
smbrc@waterboards.ca.gov
Works to ensure the long-term health of the 266-square-mile bay and its 400-square-mile watershed, which is located in the second most populous region in the United States.

Save Our Shores

www.saveourshores.org
345 Lake Ave., Ste. A
Santa Cruz, CA 95062
ph: 831-462-5660
fax: 831-462-6070
jdelay@saveourshores.org
Works to protect and conserve the marine ecosystems of
California's central coast for all generations. SOS is a nonprofit marine conservation organization with a home base
in Santa Cruz, California, and a satellite office in the Half
Moon Bay region.

Save San Francisco Bay Association

www.savesfbay.org
350 Frank H. Ogawa Plaza, Ste. 900
Oakland, CA 94612-2116
ph: 510-452-9261
fax: 510-452-9266
savebay@savesfbay.org
Works exclusively to protect and restore San Francisco Bay
and its watershed from threats of pollution and sprawl.
This is the Bay Area's oldest and largest membership
organization.

Seacology

www.seacology.org
2009 Hopkins St.
Berkeley, CA 94707
ph: 510-559-3505
fax: 510-559-3506
islands@seacology.org
Works to preserve the environments and cultures of islands
throughout the globe.

Seaflow

www.seaflow.org
1062 Fort Cronkhite
Sausalito, CA 94965
ph: 415-229-9366
fax: 415-229-9340
info@seaflow.org
Strives to protect the living oceans by working locally,
nationally, and internationally to raise awareness and promote public participation to stop harmful underwater nois-
es from military, industrial, and commercial sources that
threaten the web of life.

Sea Turtle Restoration Project

www.Seaturtles.org
PO Box 400
Forest Knolls, CA 94933
ph: 415-488-0370
fax: 415-488-0372
info@seaturtles.org
Fights to protect endangered sea turtle populations in ways
that meet the ecological needs of the sea turtles and the
oceans and the needs of the local communities who share
the beaches and waters with these creatures.

Shifting Baselines

www.shiftingbaselines.org
5254 Melrose Ave., Ste. D-112
Hollywood, CA 90038
ph: 323-960-4517
info@shiftingbaselines.org
Strives to bring attention to the severity of ocean decline.
This organization represents a media project and partnership between ocean conservation and Hollywood.

Sierra Club National Headquarters

www.sierraclub.org
85 Second St.
San Francisco, CA 94105
ph: 415-977-5500
fax: 415-977-5799
membership.services@sierraclub.org
Uses lawful means to practice and promote the responsible
use of the earth's ecosystems and resources and to educate
and enlist humanity to protect and restore the quality of
the natural and human environment. Founded in 1892 by
John Muir and other supporters, the Sierra Club has over
700,000 members and is one of the oldest conservation
organizations in the country. It has over sixty-five chapters
nationally and thirteen in California, including a number of
chapters working on coastal and ocean issues.

Sierra Club–California State Office

www.sanfranciscobay.sierraclub.org
1414 Kay St., Ste. 500
Sacramento, CA 95814
ph: 916-557-1100
fax: 916-557-9669

Sierra Club–California, Loma Prieta Chapter

www.lomaprieta.sierraclub.org
3921 East Bayshore Rd., Ste. 204
Palo Alto, CA 94303
ph: 650-390-8411
fax: 650-390-8497
loma.prieta.chapter@sierraclub.org

Sierra Club–California, Los Angeles Chapter
www.angeles.sierraclub.org
3435 Wilshire Blvd, #320
Los Angeles, CA 90010-1904
ph: 213-387-4287
fax: 213-387-5383
info@angeles.sierraclub.org

Sierra Club–California, Los Padres Chapter
www.lospadres.sierraclub.org
PO Box 90924
Santa Barbara, CA 93190
ph: 805-966-6622
los.padres.chapter@sierraclub.org

Sierra Club–California, Palos Verdes–South Bay Group
PO Box 2464
Ranchos Palos Verdes, CA 90274
ph: 310-378-3780

Sierra Club–California, Redwood Chapter
www.redwood.sierraclub.org
PO Box 466
404 Mendocino Ave., Ste. A
Santa Rosa, CA 95402
ph: 707-544-7651
fax: 707-544-9861
penningt@sonic.net

Sierra Club–California, Redwood Chapter Napa Group
www.redwood.sierraclub.org/napa
PO Box 644
Napa, CA 94559
ph: 707-966-5211
ckunze@ix.netcom.com

Sierra Club–California, Redwood Chapter North Group
www.redwood.sierraclub.org/north
PO Box 238
Arcata, CA 95518
ph: 707-445-2690
dfbeck@northcoast.com
Encompasses all Sierra Club members living in Humboldt, Del Norte, Trinity, and West Siskiyou counties.

Sierra Club–California, San Diego Chapter
www.sandiego.sierraclub.org
3820 Ray St.
San Diego, CA 92104
ph: 619-299-1744
admin@sierraclubsandiego.org

Sierra Club–California, San Francisco Chapter
www.sanfranciscobay.sierraclub.org
2530 San Pablo Ave., Ste. I
Berkeley, CA 94702-2000
ph: 510-848-0800
fax: 510-848-3383
info@sfbaysc.org
Encompasses Alameda, Contra Costa, Marin, and San Francisco counties.

Sierra Club–California, Santa Cruz Group, Ventana Chapter
www.ventana.sierraclub.org
PO Box 604
Santa Cruz, CA 95061-0604
ph: 831-426-4453
santacruz-staff@ventana.sierraclub.org

Sierra Club–California, Santa Lucia Chapter
www.santalucia.sierraclub.org
1204 Nipomo St.
San Luis Obispo, CA 93401
ph: 805-543-8717
santa.lucia.chapter@sierraclub.org

Sierra Club–California, Ventana Chapter
www.ventana.sierraclub.org
PO Box 5667
Carmel, CA 93921
ph: 831-624-8032
ventana@mbay.net
Covers Monterey and Santa Cruz counties.

Sonoma County Conservation Council
www.envirocentersoco.org
404A Mendocino Ave.
Santa Rosa, CA 95402
ph: 707-578-0595
fax: 707-544-9861
info@envirocentersoco.org
Strives to protect and enhance the coastal county's environment through the pooling of resources. A federation of local groups formed the SCCC in 1984. The SCCC sponsors the Environmental Center of Sonoma County, which is operated by volunteers.

Sonoma Land Trust
www.sonomalandtrust.org
966 Sonoma Ave.
Santa Rosa, CA 95404
ph: 707-526-6930
slt@sonic.net

Works to provide permanent protection of Sonoma County land—its natural beauty and its biotic resources—by offering stewardship, education, and guidance for the preservation and enhancement of agricultural, natural, scenic, coastal, and open space lands.

Steinhart Aquarium
www.calacademy.org/aquarium
875 Howard St.
San Francisco, CA 94103-3009
ph: 415-321-8000
info@calacademy.org
Offers exhibits featuring seabirds, fish, frogs, turtles, and other representatives of diverse water environments. The aquarium is part of the California Academy of Science.

Surfers' Environmental Alliance
www.seasurfer.org
1940 Merrill St.
Santa Cruz, CA 95060
dardley@got.net
Promotes efforts to protect the natural wonders of the coastal environment and fosters and protects beach access. SEA is a grassroots, project-based organization aggressively committed to the cultural and environmental integrity of the sport of surfing.

Surfer's Medical Association
www.damoon.net/sma
PO Box 1210
Aptos, CA 95001
SMACentral@aol.com
Strives to educate surfers so they can spend minimal time hassling with doctors and maximal time surfing. The association conducts and supports research and educational activities on surfing and health; represents the sport of surfing in the fields of medicine and science; teaches physicians about the unique health problems of surfers and how to better care for surfers; creates a network of barefoot doctors and surfing health professionals around the world; and protects and preserves the surfers' natural environment: the waves, the ocean, and the beaches.

Surfrider Foundation—National Office
www.surfrider.org
PO Box 6010
San Clemente, CA 92674-6010
ph: 800-743-SURF; 949-492-8170
info@surfrider.org
Uses conservation, activism, research, and education to encourage people to protect and enjoy the world's waves and beaches. The Surfrider Foundation is a nonprofit environmental organization whose membership consists mainly of surfers. Its scope has expanded to include protecting wetlands, bird life, and beaches as well as surf spots. It has sixty-six chapters worldwide, including twenty chapters in California.

Surfrider Foundation–California, Crescent City/Jefferson State
www.surfrider.org/crescentcity
1720 Ashford Rd.
Crescent City, CA 95531
ph: 707-458-9615

Surfrider Foundation–California, Humboldt Chapter
www.surfrider.org/humboldt
PO Box 4605
Arcata, CA 95521
ph: 707-616-5852
humboldt@surfrider.org

Surfrider Foundation–California, Huntington/Seal Beach
www.surfrider.org/huntington
PO Box 878
Huntington Beach, CA 92648
ph: 562-438-6994
info@surfrider.org

Surfrider Foundation–California, Isla Vista
www.orgs.sa.ucsb.edu/sf/
6835 Pasado
Isla Vista, CA 93117
ph: 805-685-1158
ivsurfrider@hotmail.com

Surfrider Foundation–California, Laguna Beach
www.surfrider.org/lagunabeach
668 N. Coast Hwy., #266
Laguna Beach, CA 92651
ph: 949-631-6273; 949-492-8120
laguna@surfrider.org

Surfrider Foundation–California, Long Beach
www.surfrider.org/longbeach
PO Box 14627
Long Beach, CA 90853
ph: 562-433-4323
longbeach@surfrider.org

Surfrider Foundation–California, Malibu
www.surfrider.org/SFMalibu
PO Box 953
Malibu, CA 90265-7953
ph: 310-451-1010
malibu@surfrider.org
Represents the approximately thirty-one-mile coastal environment from the Los Angeles–Ventura county line to Ballona Creek.

Surfrider Foundation–California, Marin Chapter
www.surfrider.org/marin
PO Box 1171
Larkspur, CA 94939
ph: 415-868-9445
tyeyaksb@yahoo.com

Surfrider Foundation–California, Monterey
www.surfrider.org/monterey
443 Lighthouse Ave.
Monterey, CA 93940
ph: 831-375-5015

Surfrider Foundation–California, Newport Beach
www.surfrider.org/newportbeach
323 Jasmine
Corona del Mar, CA 92625
ph: 949-631-6273
lglobalman@aol.com

Surfrider Foundation–California, South Orange County
www.surfrider.org/sanclemente
PO Box 865
San Clemente, CA 92674
ph: 949-492-8248
sanclemente@surfrider.org

Surfrider Foundation–California, San Diego
www.surfridersd.org
PO Box 1511
Solana Beach, CA 92075
ph: 858-792-9940
fax: 858-792-9940
info@surfridersd.org

Surfrider Foundation–California, San Francisco
www.sfsurfrider.org
PO Box 320146
San Francisco, CA 94132-0336
ph: 415-665-4155
sanfrancisco@surfrider.org

Surfrider Foundation–California, San Luis Bay
www.sanluisbaysurfrider.org
PO Box 3406
Pismo Beach, CA 93448
ph: 805-771-1134
info@sanluisbay.surfrider.org

Surfrider Foundation–California, San Mateo County
www.surfridersanmateoco.org
PO Box 1034
Moss Beach, CA 94304
ph: 650-728-5067
edmundo@surfridersanmateoco.org

Surfrider Foundation–California, Santa Barbara
www.rain.org/~srfrdrsb
PO Box 21703
Santa Barbara, CA 93121
ph: 805-899-BLUE
srfrdrsb@rain.org

Surfrider Foundation–California, Santa Cruz
www.surfridersantacruz.org
PO Box 3968
Santa Cruz, CA 95063
ph: 831-423-7667
surfridersantacruz@yahoo.com

Surfrider Foundation–California, Sonoma Coast
www.surfrider.org/sonomacoast
9293 Old Redwood Hwy
Penngrove, CA 94951
ph: 707-332-1083
sonoma@surfrider.org

Surfrider Foundation–California, South Bay
www.surfrider.org/southbay
PO Box 3825
Manhattan Beach, CA 90266
ph: 310-535-3116
sbaysurfrider@hotmail.com

Surfrider Foundation–California, Ventura
www.surfrider.org/Ventura
239 W. Main St.
Ventura, CA 93001
ph: 805-667-2222
paul@matilija-coalition.org

Sustainable Fishery Advocates
www.sustainablefishery.org
PO Box 233
Santa Cruz, CA 95062
ph: 831-427-1707
fax: 309-213-4688
rboot@sustainablefishery.org
Partners with markets, restaurants, and distributors to create markets for sustainably harvested seafood products. Through education and a flexible labeling system, the organization seeks to decrease unsustainable fishing practices while improving the livelihoods of people who fish, fish populations, and ocean ecosystems.

30 Minute Beach Cleanup
www.beachcleanup.org
1 Granada Ave.
Long Beach, CA 90803
ph: 562-439-2681
justinrudd@aol.com
Helps keep beaches safe and clean in Long Beach.

Tomales Bay Institute
www.earthisland.org/tbi/index.cfm
www.onthecommons.org
PO Box 127
Point Reyes Station, CA 94956
ph: 415-663-8560
jonrowe@tomales.org
Seeks to expand the scope of the possible in American poli-
tics and policy and protect common resources including the
global ocean commons. The institute was founded in 2001.

Trout Unlimited–California, State Office
www.tucalifornia.org
1120 College Ave.
Santa Rosa, CA 95404
ph: 707-543-5877
fax: 707-543-5857
info@tucalifornia.org
Works to conserve, protect, and restore North America's
trout and salmon fisheries and their watersheds.

Trout Unlimited–California, Council
www.tucalifornia.org
828 San Pablo Ave. #208
Albany, CA 94706-1678
ph: 510-528-5390
fax: 510-528-7880
tucalif@earthlink.net

United Anglers of California
www.unitedanglers.org
15572 Woodard Rd.
San Jose, CA 95124
ph: 408-371-0331
bobstrickland@unitedanglers.org
Endeavors to stop the decline of fisheries by ensuring
resources are properly managed. Founded in 1981, the
organization's goal is to unite anglers, fishing groups, businesses, and citizens into a powerful statewide fishery conservation organization.

United Anglers of Southern California
www.unitedanglers.com
5948 Warner Ave.
Huntington Beach, CA 92649
ph: 714-840-0227
fax: 714-840-3146
info@unitedanglers.com
Works to preserve the marine environment. The not-for-
profit organization made up of volunteers recognizes the
need for commercial fisheries but opposes the indiscrimi-
nate depletion of the resource for short-term profit.

Ventana Wilderness Society
www.ventanaws.org
19045 Portola Dr., Ste. F-1
Salinas, CA 93908
ph: 877-897-7740
fax: 831-455-2846
info@ventanaws.org
Perpetuates plant and animal species native to the central
California coast through wildlife and habitat reintroduction,
restoration, research, and education.

Ventura Coastkeeper/Wishtoyo Foundation
www.wishtoyo.org
3600 South Harbor Blvd., Ste. 222
Oxnard, CA 93035
ph: 805-382-4540
fax: 805-382-4541
info@wishtoyo.org
Preserves the wisdom of the Chumash culture and links it
to present-day environmental issues, including protection of
Ventura County's marine habitat, coastal waters, and
watershed. The Ventura Coastkeeper is the first Native
American group to become a keeper.

Ventura County Maritime Museum
2731 S. Victoria Ave.
Oxnard, CA 93035
ph: 805-984-6260
vcmm@aol.com
Offers comprehensive maritime exhibits including a world-
class collection of historic ship models and fine-art maritime paintings.

Vote the Coast
www.votethecoast.org
PO Box 1022
Malibu, CA 90265
ph: 310-456-5674
executivedirector@votethecoast.org
Works to achieve coastal protection and conservation
through knowledge, public education, a strategic assertion
of power, and forming alliances to develop a coordinated
policy of coastal protection. A grassroots political action
organization, Vote the Coast was created in 1996 to educate voters on the importance of the protection of
California's fragile coastal resources and let them know
how their vote affects coastal protection.

Wilderness Conservancy
www.wildcon.org
1224 Roberto Ln.
Los Angeles, CA 90077-2334
ph: 310-472-2593
info@wildcon.org
Works to conserve endangered and threatened wildlife and

wild places on land or in the sea. The conservancy is a direct-action foundation that provides assets to people and organizations (governmental and nongovernmental) with specific needs that cannot be addressed because of lack of funding.

Windows on Our Waters
www.windowsonourwaters.org
2515 Wilshire Blvd.
Santa Monica, CA 90403
ph: 310-829-1229 x235
garypoe@windowsonourwaters.org
Offers teaching about the issues of nonpoint source (NPS) pollution and urban runoff and their effects on humans as well as coastal and marine environments.

Yurok Tribe
190 Klamath Blvd.
PO Box 1027
Klamath, CA 95548
ph: 707-482-1350
fax: 707-482-1377
Works to protect wild salmon and other sacred fish of the Klamath River and defend the tribe's treaty rights from government water diversions. The Yurok Tribe is largest Indian nation in the state of California, with nearly 5,000 members.

COLORADO
Colorado's Ocean Journey
www.oceanjourney.org
700 Water St. Qwest Park
Denver, CO 80211
ph: 303-561-4450
fax: 303-561-4465
guestinfo@oceanjourney.org
Strives to positively impact the natural environment by providing leadership and resources for education, preservation, and conservation.

CONNECTICUT
Connecticut Audubon Society Coastal Center
www.ctaudubon.org/centers/coastal/costal.htm
1 Milford Point Rd.
Milford, CT 06460
ph: 203-878-7440
fax: 203-876-2813
Connecticut Audubon opened its Coastal Center in 1995. The center is located on an 8.4-acre barrier beach situated next to the 840-acre Wheeler Wildlife Management Area (a salt marsh) at the mouth of the Housatonic River.

Connecticut Audubon Society Center at Fairfield
www.ctaudubon.org
2325 Burr St.
Fairfield, CT 06824
ph: 203-259-6305
fax: 203-254-7673
dryan@ctaudbon.org
Works to conserve and restore natural ecosystems by focusing on birds, other wildlife, and their habitats for the benefit of humanity and the earth's biological diversity. Audubon is a national network of community-based nature centers and chapters, scientific and educational programs, and advocacy on behalf of areas sustaining important bird populations. Connecticut Audubon began as a small group of concerned citizens organized to protect birds at the end of the nineteenth century. It is now a statewide organization focusing on environmental education.

Cetacean Society International
www.csiwhalesalive.org
PO Box 953
Georgetown, CT 06829
ph: 203-770-8615
fax: 860-561-0187
rossiter@csiwhalesalive.org
Works to achieve on a global basis the optimum utilization of cetacean resources through benign use and the elimination of all killing and captive display of whales, dolphins, and porpoises. CSI is an all-volunteer, nonprofit conservation, education, and research organization based in the United States, with volunteer representatives in twenty-six countries around the world.

Clean Sound Inc.
www.cleansound.org
20 Ojibwa Rd.
Shelton, CT 06484
info@cleansound.org
Works to remove fixed and floatable pollutants from Long Island Sound and all contributing watershed areas, to restore degraded habitats, and to inform others about the pollution problems facing the Long Island Sound. Established in 1990, Clean Sound provides the opportunity for people to volunteer in a hands-on solution to the pollution problems facing Long Island Sound.

Coast Guard Academy
www.cga.edu
31 Mohegan Ave.
New London, CT 06320-8103
ph: 860-444-8444
Trains officers for service in the U.S. Coast Guard, which is responsible for patrolling and protecting U.S. coastal waters.

Connecticut Fund for the Environment
www.cfenv.org
205 Whitney Ave.
New Haven, CT 06511
ph: 203-787-0646
fax: 203-787-0246
protect@cfenv.org
Uses law, science, and education to improve air and water
quality, control toxic contamination, minimize the adverse
impacts of highways and traffic congestion, protect public
water supplies, and preserve the open space and wetlands
crucial to both the state's citizens and its wildlife. CFE was
founded in 1978.

Connecticut PIRG
www.connpirg.org
198 Park Rd., 2nd Fl.
West Hartford, CT 06119
ph: 860-233-7554
fax: 860-233-7574
info@connpirg.org
Educates citizens to participate in the political process and
works on Connecticut coastal issues.

Environmental Defense–New England
www.environmentaldefense.org
368 Noank Rd.
Mystic, CT 06355
ph: 860-572-0190
fax: 860-371-3701
Works to restore public fisheries resources and manage
species that depend on healthy marine ecosystems. Sally
McGee, who sits on the New England Regional Fishery
Management Council, is responsible for most of this work.

Long Island Sound Foundation
www.lisfoundation.org
1080 Shennecossett Rd.
Groton, CT 06340
ph: 860-405-9166
mail@lisfoundation.org
Facilitates the exchange of information among individuals
and organizations and enhances their ability to address
issues impacting Long Island Sound; enhances public learning, awareness, understanding, and involvement focused on
Long Island Sound. The foundation also established an
environmental fund to support scientific and public policy
research, education, and community programs.

Long Island Soundkeeper
www.soundkeeper.org
PO Box 4058
Norwalk, CT 06855
ph: 203-854-5330
fax: 203-866-1318

info@soundkeeper.org
Works to protect and enhance the biological, physical, and
chemical integrity of Long Island Sound and its watershed
through patrolling, investigating, intervening, and raising
public awareness of the Sound's problems.

The Maritime Aquarium at Norwalk
www.maritimeaquarium.org
10 North Water St.
Norwalk, CT 06854
ph: 203-852-0700
fax: 203-838-5416
Seeks to ignite curiosity and promote understanding of all
living creatures, especially those around Long Island Sound
and its watershed.

Mystic Marinelife Aquarium
www.mysticaquarium.org
55 Coogan Blvd.
Mystic, CT 06355-1997
ph: 860-572-5955
info@mysticaquarium.org
Offers marine wildlife exhibits and interactive programs
including exhibits on the *Titanic,* dolphins, and ancient
seas. The aquarium is also home to Dr. Robert Ballard's
Institute for Exploration.

Office of Long Island Sound Programs
http://dep.state.ct.us/olisp
79 Elm St., 3rd Fl.
Hartford, CT 06106-5127
ph: 860-424-3034
fax: 860-424-4054
dep.webmaster@po.state.ct.us
Protects, manages, and restores coastal resources and
ensures their availability and accessibility to the public. This
state agency also fosters water-dependent uses of the
shorefront and oversees the state's public trust responsibilities for tidelands.

Save the Sound
www.savethesound.org
18 Reynolds St.
East Norwalk, CT 06855
ph: 203-354-0036
fax: 203-354-0036
savethesound@savethesound.org
Works on restoration, protection, and appreciation of Long
Island Sound and its watershed through advocacy, education, and research. Save the Sound is a bi-state, nonprofit
membership organization.

Schooner Incorporated
www.schoonerinc.org
60 South Water St.
New Haven, CT 06519
ph: 203-865-1737
fax: 203-624-8816
director@schoonerinc.org
Promotes conservation of the environment, particularly
Long Island Sound and the rivers of Connecticut, and
appreciation of their culture, history, and future.

DELAWARE
Delaware Center for the Inland Bays
www.inlandbays.org
16529 Coastal Hwy.
Lewes, DE 19958
ph: 302-645-7325
fax: 302-645-5765
information@inlandbays.org
Works to restore and protect three interconnected waterways in southeastern Sussex County through research,
restoration work, and educational programs.

Green Delaware
www.greendel.org
PO Box 69
Port Penn, DE 19731
ph: 302-834-3466
fax: 302-836-3005
greendel@dca.net
Advocates policies consistent with good health, preservation of biodiversity, and long-term sustainability. This is a
grassroots organization concerned with environmental and
public health issues in Delaware and surrounding states
that has participated in marine conservation efforts around
clean-water discharge issues and coastal sprawl.

Partnership for the Delaware Estuary
www.greenworks.tv/delawareestuary
One Riverwalk Plaza
110 S. Poplar St., Ste. 202
Wilmington, DE 19801
ph: 800-445-4935; 302-655-4990
fax: 302-655-4991
partners@udel.edu
Supports conservation of the Delaware Estuary by promoting education on the estuary and coordinating habitat
restoration projects for corporations. Through a minigrant
program, the partnership provides funding and support for
other nonprofits.

Sierra Club–Delaware Chapter
www.delaware.sierraclub.org
100 West 10th St., Ste. 1107
Wilmington, DE 19801
ph: 302-425-4911
delaware.chapter@sierraclub.org
Advocates exploring, enjoying, and protecting the planet,
as well as fighting pollution, exploring area wetlands, and
educating the public through programs such as the coastal
hybrid evolution tour for cleaner energy.

Surfrider Foundation–Delaware
www.surfrider.org/delaware
1000 Canterbury Rd.
Milford, DE 19963
ph: 302-645-0287
yakermtn@earthlink.net
Uses conservation, activism, research, and education to
encourage people to protect and enjoy the world's waves
and beaches. The Surfrider Foundation is a nonprofit environmental organization whose membership consists mainly
of surfers. Its scope has expanded to include protecting
wetlands, bird life, and beaches as well as surf spots like
those along the Delaware shore.

DISTRICT OF COLUMBIA
American Rivers
www.amrivers.org
1025 Vermont Ave. NW, Ste. 720
Washington, DC 20005-3516
ph: 202-347-7550
alevine@mail.mcn.org
Works to protect and restore healthy, natural rivers and the
variety of life they sustain for people, fish, and wildlife.

Atlantic CoastWatch
www.atlanticcoastwatch.org
3121 South St. NW
Washington, DC 20007
ph: 202-338-1017
susdev@igc.org
Publishes a free newsletter about marine conservation
efforts taking place in the region between the Caribbean
and Nova Scotia.

Audubon Society–District of Columbia Chapter
www.dcaudubon.org
PO Box 15726
1150 Connecticut Ave. NW, Ste. 600
Washington, DC 20036
ph: 202-861-2242
mikekaspar@aol.com
Works to conserve and restore natural ecosystems, focusing
on birds, other wildlife, and their habitats for the benefit of
humanity and the earth's biological diversity. Audubon is a
national network of community-based nature centers and
chapters, scientific and educational programs, and advocacy
on behalf of areas sustaining important bird populations.

Blue Frontier Campaign
www.bluefront.org
PO Box 19367
Washington, DC 20036
ph: 202-387-8030
fax: 202-234-5176
info@bluefront.org
Works to strengthen America's ocean constituency by building unity among seaweed (marine grassroots) activists at the local, regional, and national levels. Provides organizing tools such as books, conferences, field trainings, and awards for ocean heroes and works to heighten public awareness about our ocean frontier and practical solutions for its enjoyment, protection, and restoration.

Center for Food Safety
www.centerforfoodsafety.org
660 Pennsylvania Ave. SE, Ste. 302
Washington, DC 20003
ph: 202-547-9359
fax: 202-547-9429
office@centerforfoodsafety.org
Challenges harmful food production technologies and promotes sustainable alternatives, including in the harvesting and consumption of seafood. CFS is a nonprofit public interest and environmental advocacy membership organization.

Clean Beaches Council
www.cleanbeaches.org
1225 New York Ave. NW, Ste. 450
Washington, DC 20005
ph: 202-682-9507
fax: 202-682-9506
info@cleanbeaches.org
Promotes public awareness and volunteer participation in sustainability while ensuring a legacy of clean beaches for all generations to come.

Clean Water Fund
www.cleanwaterfund.org
4455 Connecticut Ave. NW, Ste. A300-16
Washington, DC 20008-2328
ph: 202-895-0432
fax: 202-895-0438
cwf@cleanwater.org
Helps people campaign successfully for cleaner and safer water, cleaner air, and protection from toxic pollution in homes, neighborhoods, and workplaces. CWF was founded in 1978.

Clean Water Network
www.cwn.org
1200 New York Ave. NW, Ste. 400
Washington, DC 20005
ph: 202-289-2395
fax: 202-289-1060
ksmitherman@nrdc.org
Works to strengthen and implement federal clean water and wetlands policy. CWN is an alliance of more than a thousand public interest organizations around the country.

Coastal America
www.coastalamerica.gov
300 7th St. SW, Ste. 680
Washington, DC 20250-0599
ph: 202-401-9928
fax: 202-401-9821
darlenemaphis@usda.gov
Works to protect, preserve, and restore the nation's coasts. Coastal America is a partnership of federal agencies, state and local governments, and private organizations.

Coast Alliance
www.coastalliance.org
Presently in the process of moving, the Coast Alliance will have contact information posted on its Web site.
coast@coastalliance.org
Concerned about the developmental pressure and pollution along the coasts, the alliance is a marine grassroots (seaweed) network of citizen-activist organizations developing viable alternatives for sustainable coasts and communities.

Coastal States Organization
www.sso.org/cso
444 North Capitol St. NW
Washington, DC 20001
ph: 202-508-3860
cso@sso.org
Advocates for improved management of the nation's coasts and Great Lakes. CSO represents the interests of the states in carrying out the Coastal Zone Management Act and in other aspects of federal policy that might impact the protection and sustainable development of coastal states.

Conservation International–Marine Program
www.conservation.org/xp/CIWEB/strategies/
marine_ecosystems/marine
1919 M St. NW, Ste. 600
Washington, DC 20036
ph: 800-406-2306; 202-912-1000
Applies innovations in science, economics, policy, and community participation to protect the earth's richest regions of plant and animal diversity in the hot spots, major tropical wilderness areas, and key marine ecosystems.

C.O.R.E. (Consortium for Oceanographic Research and Education)
www.coreocean.org
1201 New York Ave. NW, Ste. 420
Washington, DC 20005
ph: 202-332-0063
fax: 202-332-8887
core@COREocean.org
Promotes a number of programs to encourage coordination on ocean research between federal, private, and academic interests, to carry out long-term studies of the ocean, and to increase scientific literacy among students and the public. C.O.R.E. is an association of U.S. oceanographic research institutions and aquaria. Its seventy-nine members represent the nucleus of U.S. research and education about the ocean.

Defenders of Wildlife
www.defenders.org
1130 17th St. NW
Washington, DC 20036
ph: 202-682-9400
fax: 202-682-1331
info@defenders.org
Works to protect all native wild animals and plants in their natural communities, including marine plants and animals in coastal and offshore habitats.

EarthEcho International
www.Earthecho.org
1050 Connecticut Ave. NW, Ste. 1000
Washington, DC 20036
ph: 202-772-4272
fax: 202-772-3101
philippe@earthecho.org
Strives to accomplish a measurable reduction in the damage to and loss of natural resources and habitat among the world's oceans. EarthEcho works to inspire a sense of stewardship for the planet and create an environmentally sustainable future. Founded by Philippe and Alexandra Cousteau.

Earthjustice Legal Defense Fund–District of Columbia
www.earthjustice.org/regional/washington_dc
1625 Massachusetts Ave. NW, Ste. 702
Washington, DC 20036
ph: 202-667-4500
eajusdc@earthjustice.org
Shapes the development of environmental law and actively participates in a number of battles to protect coastal and marine resources. Earthjustice was founded in 1971 as Sierra Club Legal Defense Fund.

Ecological Society of America
www.esa.org
1707 H St. NW, Ste. 400
Washington, DC 20006
ph: 202-833-8773
fax: 202-833-8775
esahq@esa.org
Promotes ecological science by improving communication among ecologists and raising the public's level of awareness of the importance of ecological science. The society was founded in 1915.

Environmental Defense–District of Columbia
www.environmentaldefense.org
Regional Office
1875 Connecticut Ave. NW
Washington, DC 20009
ph: 202-387-3500
fax: 202-234-6049
members@environmentaldefense.org
Links science, economics, and law to create innovative, equitable, and cost-effective solutions to society's most urgent environmental problems. The DC office is also home to the organization's oceans program.

Environmental Law Institute
www.eli.org
2000 L St. NW, Ste. 620
Washington, DC 20036
ph: 202-939-3800
fax: 202-939-3868
law@eli.org
Advances environmental protection by improving law, policy, and management, with recent interest in protection of the public trust doctrine as it applies to America's ocean commons.

Environmental Literacy Council
www.enviroliteracy.org
1625 K. St. NW, Ste. 1020
Washington, DC 20006
ph: 202-296-0390
fax: 202-822-0991
info@enviroliteracy.org
Gives teachers tools to help students develop environmental literacy, that is, a fundamental understanding of the systems of the world, both living and nonliving. The council is an independent, nonprofit organization.

Greenpeace USA–District of Columbia
www.greenpeaceusa.org
702 H St. NW, Ste. 300
Washington, DC 20001
ph: 800-326-0959; 202-319-2498
fax: 202-462-4507

Uses nonviolent direct action and creative communication to expose global environmental problems and to promote solutions that are essential to a green (blue) and peaceful future. Greenpeace is an international independent membership organization whose goal is to ensure the ability of Earth to nurture life in all its diversity. Its present campaigns include efforts to protect ocean seamounts and eliminated destructive bottom-trawl fishing.

House Oceans Caucus

www.house.gov/greenwood/OCEAN/hoc_mission.html
2466 Rayburn HOB
Washington, DC 20515
Builds awareness, exchanges information and ideas, explores important issues, and develops and implements ocean policy legislation to meet national and international needs. The House Oceans Caucus is a bipartisan forum made up of members of the U.S. House of Representatives.

Island Resources Foundation

www.irf.org
1718 P St. NW, Ste. T-4
Washington, DC 20036
ph: 202-265-9712
fax: 202-232-0748
bpotter@irf.org
Provides development and environmental planning assistance to governments and private nonprofit environmental organizations of small tropical islands. This organization is an educational and research foundation.

John Heinz Center for Science, Economics and the Environment

www.heinzctr.org
1001 Pennsylvania Ave. NW, Ste. 735 S.
Washington, DC 20004
ph: 202-737-6307
fax: 202-737-6410
Strives to improving the scientific and economic foundation for environmental policy through multisectoral collaboration among industry, government, academia, and environmental organizations.

League of Conservation Voters

www.lcv.org
1920 L St. NW, Ste. 800
Washington, DC 20036
ph: 202-785-8683
fax: 202-835-0491
Works full time to shape a pro-environment Congress and White House.

Marina Industries (Association of)

www.marinaassociation.org
444 N. Capitol St. NW, Ste. 645
Washington, DC 20001
ph: 202-737-9776
fax: 202-628-8679
info@marinaassociation.org
Represents the marine industry on a range of regulatory and legislative issues. MOAA is the national trade association of the marina industry. Its membership includes some 950 marinas, boatyards, and yacht clubs.

Marine Conservation Biology Institute– Public Policy Office

www.mcbi.org
600 Pennsylvania Ave. SE, Ste. 210
Washington, DC 20003
ph: 202-546-5346
fax: 202-546-5348
mcbiweb@mcbi.org
Works to protect and restore marine life on the West Coast, around the United States and beyond, through research and training in marine conservation biology. The DC office functions as the public policy branch of MCBI.

Marine Fish Conservation Network

www.conservefish.org
600 Pennsylvania Ave. SE, Ste. 210
Washington, DC 20003
ph: 866-823-8552; 202-543-5509
fax: 202-543-5774
network@conservefish.org
Works to conserve marine fish and promotes their long-term sustainability. The network is a coalition of over 160 national and regional environmental organizations, commercial and recreational fishing groups, aquariums, and marine science groups.

Marine PhotoBank

www.marinephotobank.org
1731 Connecticut Ave. NW, 4th Floor
Washington, DC 20009
ph: 202-483-9570
marinephotobank@seaweb.org
Offers a venue for photographers to contribute high-quality images of threats to our marine world, enabling nongovernmental organizations and government agencies to use the pictures, royalty free, to educate the public about the ocean's needs and join the effort to protect the water world.

National Aquarium
www.nationalaquarium.com
Department of Commerce Building
14th and Constitution Ave. NW
Washington, DC 20230
ph: 202-482-2825
fax: 202-482-4946
info@nationalaquarium.com
Features more than 80 exhibits with over 200 species of
salt water and freshwater marine life including endangered
species. The theme of the nation's oldest aquarium is
America's marine sanctuaries.

National Environmental Trust
www.environet.policy.net/marine
1200 Eighteenth St. NW, 5th Fl.
Washington, DC 20036
ph: 202-887-8800
fax: 202-887-8877
netinfo@environet.org
Informs citizens about environmental problems and their
effects on health and quality of life. Ventures include a
major ocean protection project focusing on marine fisheries
reform.

National Fish and Wildlife Foundation
www.nfwf.org/index.htm
1120 Connecticut Ave. NW, Ste. 900
Washington, DC 20036
ph: 202-857-0166
fax: 202-857-0162
pico@nfwf.org
Conserves healthy populations of fish, wildlife, and plants,
on land and in the sea, through creative and respectful
partnerships, sustainable solutions, and better education.

National Geographic Society
www.nationalgeographic.com/index.html
1145 17th St. NW
Washington, DC 20036-4688
ph: 800-647-5463
askngs@nationalgeographic.com
Promotes concern about the alarming lack of geographic
knowledge among the nation's young people and the
pressing need to protect the planet's natural resources.
Sponsors a number of marine exploration and science projects; also does extensive media reporting on marine discoveries and conservation issues.

National Wildlife Federation–District of Columbia
www.nwf.org/about/
1400 16th St. NW, Ste. 501
Washington, DC 20036
ph: 202-797-6800
fax: 202-797-5846
Educates and empowers people from all walks of life to
protect wildlife and habitat for future generations by providing hands-on conservation education and action
opportunities.

**Natural Resources Defense Council–
District of Columbia Regional Office**
www.nrdc.org/default.asp
1200 New York Ave. NW, Ste. 400
Washington, DC 20005
ph: 202-289-6868
nrdcinfo@nrdc.org
Uses law, science, and the support of more than one million
online activists to protect the planet's wildlife and wild
places, and to ensure a safe and healthy environment for
all living things. NRDC also has an active marine program.

Navy Museum
www.history.navy.mil/branches/nhcorg8.htm
Washington Navy Yard, Building 76
805 Kidder Breese SE
Washington, DC 20374-5060
ph: 202-433-4882
fax: 202-433-8200
archives@navy.mil
Houses displays and artifacts on naval history including the
research submersible *Trieste*, the only craft ever to take
humans to the deepest part of the world's oceans, the
Challenger Deep in the Marianas Trench. Since the terrorist
attacks of 9/11, members of the public must request a special security pass to visit this site.

**North American Association for Environmental
Education**
www.naaee.org
2000 P St. NW, Ste. 540
Washington, DC 20036
ph: 202-419-0412
fax: 202-419-0415
email@naaee.org
Promotes and supports environmental education and
educators through a network of professionals, students,
and volunteers working in environmental education
throughout North America and in over fifty-five countries
around the world.

Oceana
www.oceana.org
2501 M St. NW, Ste. 300
Washington, DC 20037-1311
ph: 877-7-OCEANA; 800-8-OCEAN-0; 202-833-3900
fax: 202-833-2070
pkline@oceana.org
Campaigns to protect and restore the world's oceans by
using teams of marine scientists, economists, lawyers, and
advocates to help enact specific and concrete policy
changes to reduce pollution and to prevent the collapse of
fish populations, marine mammals, and other sea life.

The Ocean Conservancy–District of Columbia
www.oceanconservancy.org
2029 K St. NW
Washington, DC 20006
ph: 800-519-1541; 202-429-5609
fax: 202-872-0619
info@oceanconservancy.org
Uses science-based advocacy, research, and public education to inform, inspire, and empower people to speak and
act for the oceans. The Ocean Conservancy is a national
organization dedicated to protecting ocean ecosystems.
Its major programs include coral reef protection, debris
monitoring, stormwater runoff and pollution, and coastal
cleanup.

The Ocean Studies Board
www.nationalacademies.org/osb
500 5th St., Rm. 752
Washington, DC 20001
ph: 202-334-2714
fax: 202-334-2885
Advises the federal government on issues of ocean science,
engineering, and policy. In addition to exercising leadership
within the ocean community, the board undertakes studies
that have often raised concerns about major environmental
and other problems confronting the public seas. The
National Research Council established this board.

1Planet1Ocean
www.1planet1ocean.org
PO Box 53090
Washington, DC 20009
ph: 309-216-5870
fax: 309-216-5870
info@1planet1ocean.org
Seeks to explore, restore, and sustain the oceans through
strong international partnerships. The organization's first
major project seeks to make the Gulf of Mexico a model
of sustainability.

Public Employees for Environmental Responsibility
www.peer.org
2001 S St. NW, Ste. 570
Washington, DC 20009
ph: 202-265-7337
fax: 202-265-4192
info@peer.org
Promotes responsible management of America's public
resources including our seas. PEER has helped expose and
publicize wrongdoing and defended employees from a
number of marine-related agencies, including the U.S. Fish
and Wildlife Service and the National Oceanic and
Atmospheric Administration. PEER is a national alliance of
local, state, and federal resource professionals working
nationwide with government scientists, land managers,
environmental law enforcement agents, field specialists,
and other resource professionals.

River Network
www.rivernetwork.org
3814 Albemarle St. NW
Washington, DC 20016
ph: 202-364-2550
fax: 202-364-2520
pmunoz@rivernetwork.org
Promotes caring for rivers for by those who use them and
live in their watershed. The network comprises grassroots
river and watershed conservation organizations, public
agencies, tribal governments, and coalitions working to
save freshwater ecosystems.

SeaWeb
www.seaweb.org
1731 Connecticut Ave. NW, 4th Fl.
Washington, DC 20009
ph: 202-483-9570
seaerb@seaweb.org
Raises awareness and inspires action to protect the ocean
and the life within it. Established as a media promoter for
the oceans, Seaweb has generated a number of spin-off
campaigns on sustainable seafood, conservation education
in public aquariums, and public awareness about the problems of aquaculture.

Smithsonian Institution
www.si.edu
1000 Jefferson Dr. SW
Washington, DC 20560
ph: 202-357-2700; 202-357-2627
info@si.edu
Provides authoritative experiences that connect America's
history and heritage by promoting innovation, research, and
discovery in science. The Smithsonian has a number of
marine research programs and facilities and is building a
$60 million ocean section in its Museum of Natural History.

Sprawl Watch Clearinghouse
www.sprawlwatch.org
1100 17th St. NW, 10th Fl.
Washington, DC 20036
Identifies and disseminates information on the best land-
use practices, which can help reduce polluted runoff and
other marine problems generated in coastal areas that
make up the seventeen of the twenty fastest growing
U.S. counties.

Surfrider Foundation–District of Columbia Chapter
4900 Battery Ln., #402
Bethesda, MD 20814
Uses conservation, activism, research, and education to
encourage people to protect and enjoy the world's waves
and beaches. The Surfrider Foundation is a nonprofit environmental organization whose membership consists mainly
of surfers. Its scope has expanded to include protecting
wetlands, bird life, and beaches as well as surf spots. It has
sixty-six chapters worldwide. The DC chapter members,
when not working for ocean conservation, surf Maryland,
Delaware, and the Outer Banks of North Carolina.

Taxpayers for Common $ense
www.taxpayer.net
651 Pennsylvania Ave. SE, 2nd Fl.
Washington, DC 20003
ph: 800-829-7293; 202-546-8500
fax: 202-546-8511
Promotes cutting wasteful government spending and subsidies to achieve responsible and efficient government that
lives within its means. The organization has identified a
number of federal subsidies that wastefully impact the
coasts and oceans including parts of the beach replenishment, dredging, and federal flood insurance programs.

**U.S. Public Interest Research Group–
District of Columbia**
www.uspirg.org
218 D St. SE
Washington, DC 20003
ph: 202-546-9707
fax: 202-546-2461
uspirg@pirg.org
Uses investigative research, media exposés, grassroots
organizing, advocacy, and litigation to uncover threats to
public health and well-being and fight to end them. PIRG
also has established an ocean advocacy team.

Women's Aquatic Network Inc.
www.womensaquatic.net
PO Box 4993
Washington, DC 20008
info@womensaquatic.net
Brings together women and men in the District of Columbia
with interests in marine and aquatic policy, research, and
legislation.

World Resources Institute
www.wri.org
10 G St. NE, Ste. 800
Washington, DC 20002
ph: 202-729-7600
fax: 202-729-7601
front@wri.org
Works to protect the earth and improve people's lives. The
institute is an independent organization with a staff of scientists, economists, policy experts, business analysts, statis-
tical analysts, mapmakers, and communicators. Among the
research topics they cover are Coastal and Marine
Ecosystems. They have developed a map-based indicator of
reefs at risk in different parts of the world.

World Wildlife Fund
www.worldwildlife.org
1250 24th St. NW, Ste. 500
Washington, DC 20037
ph: 202-293-4800
Directs conservation efforts toward three global goals: saving endangered species; protecting endangered habitats;
and addressing global threats such as toxic pollution, overfishing, and climate change.

FLORIDA
American Littoral Society–Southeast Chapter
www.sealitsoc.org
4154 Keats Dr.
Sarasota, FL 34241
ph: 941-377-5459
bullochd@quickenet.com
Promotes the environmental well-being of coastal habitat.
The Southeast chapter is an all-volunteer group.

Apalachicola Riverkeeper
www.abark.org
23 Avenue D
Apalachicola, FL 32320
ph: 850-653-8936
riverkeeper@abark.org
Works to protect and preserve Florida's most endangered
river and estuary—the Apalachicola—including its
tributaries. This is a grassroots environmental advocacy
organization.

Aquarius Underwater Lab NURC–UNC

www.uncw.edu/aquarius
515 Caribbean Dr.
Key Largo, FL 33037
ph: 305-451-0233
fax: 305-453-9719
otto@nurc.net
Provides the opportunity for scientists to spend up to eight days at a time living and working on the reef. This facility, run by the National Undersea Research Center, is the last underwater laboratory in the world.

Audubon of Florida

www.audubonofflorida.org
444 Brickell Ave., Ste. 850
Miami, FL 33131
ph: 305-371-6399
fax: 305.371.6398
Works to conserve and restore natural ecosystems and to promote wise stewardship of natural resources, including marine resources. Audubon is a national network of community-based nature centers and chapters, scientific and educational programs, and advocacy on behalf of areas sustaining important bird populations.

Big Pine Key Civic Association

Box 190
Big Pine Key, FL 33043
Works to protect Florida's keys and wetlands through grassroots citizen action.

The Billfish Foundation

www.billfish.org
2161 E. Commercial Blvd., 2nd Fl.
Ft. Lauderdale, FL 33308
ph: 800-438-8247; 954-938-0150
fax: 954-938-5311
tbf@billfish.org
Supports conservation and enhancement of billfish populations worldwide through scientific research, education, and advocacy.

Biscayne Bay Foundation

2964 Aviation Ave., Ste. 300
Coconut Grove, FL 33133
ph: 305-447-4566
fax: 305-447-4566
Promotes preservation and protection of Biscayne Bay and Biscayne National Park. The foundation is a nonprofit and community service organization.

Caribbean Conservation Corporation and Sea Turtle Survival League

www.cccturtle.org
4424 NW 13th, Ste. A-1
Gainesville, FL 32609
ph: 800-678-7853; 352-373-6441
fax: 352-375-2449
ccc@cccturtle.org
Works to improve the survival outlook for several species of sea turtles. Founded in 1959, CCC is a world-renowned leader in sea turtle research and conservation.

Caribbean Marine Research Center

www.perryinstitute.org
250 Tequesta Dr., Ste. 304
Tequesta, FL 33469
ph: 561-741-0192
fax: 561-741-0193
pims@perryinstitute.org
Improves and enhances understanding of the wider Caribbean region's marine environment by supporting and conducting high quality marine research and education programs.

Choctawhatchee Basin Alliance

www.basinalliance.org
100 College Blvd.
Niceville, FL 32578
ph: 850-650-9330
cba@owc.edu
Promotes optimum use of water resources to preserve environmental quality, the quality of public life, and sustain future utilization of those resources.

Clearwater Marine Aquarium

www.CMAquarium.org
249 Windward Passage
Clearwater, FL 33767-2244
ph: 888-239-9414; 727-441-1790
MLA@CMAquarium.org
Promotes public education; marine research; animal-assisted therapy; and the rescue, rehabilitation, and release of sick or injured marine animals.

Coastal Conservation Association–Florida

www.ccaflorida.org
3333 S. Orange Ave., Ste. 103
Orlando, FL 32806
ph: 407-854-7002
fax: 407-854-1766
info@ccaflorida.org

Works to protect the state's marine resources and the interests of saltwater anglers by supporting law enforcement, reasonable access for family recreational fishing, and fairly balanced fishery regulations to protect state and federal fish stocks.

Common Cause Florida

www.commoncause.org
704 W Madison St.
Tallahassee, FL 32304
ph: 850-222-3883
fax: 850-222-3906
cmncause@infionline.net
Lobbies the U.S. Congress through active, well-informed citizens organized by congressional districts. Common Cause Florida has addressed a number of coastal protection issues.

Conservancy of Southwest Florida

www.conservancy.org
1450 Merrihue Dr.
Naples, FL 34102
ph: 239-262-0304
fax: 239-262-0672
info@conservancy.org
Uses advocacy, education, and stewardship to acquire and protect critical environmental areas through land acquisition and to achieve environmentally responsible solutions to growth and development.

Coral Reef Research and Rescue Society

www.Reefsavers.org
PO Box 510461
Melbourne Beach, FL 32951
ph: 321-409-9857
divingforlife@cfl.rr.com
Directs efforts toward saving the coral reefs of Florida. A current priority is monitoring and tracking an invasive species of algae. CRRRS also promotes education about ocean-related issues, and teaching children about the environment and ways they can prevent damage to coral reefs.

Crocodile Lake National Wildlife Refuge

PO Box 370
Key Largo, FL 33037
ph: 305-451-4223
keydeer@fws.gov
Protects critical breeding and nesting habitat for the endangered American crocodile and other wildlife.

Cry of the Water

www.cryofthewater.org
PO Box 8143
Coral Springs, FL 33075
ph: 954-753-9737
reefteam2@yahoo.com

Works to protect coral reefs, including some of Florida's most northerly living reefs off Fort Lauderdale. Cry is a Reefkeeper International affiliate and a Reef Check member.

Dolphin Research Center

www.dolphins.org
58901 Overseas Highway
Grassy Key, FL 33050-6019
ph: 305-289-1121
fax: 305-743-7627
drc@dolphins.org
Studies interactions between the public and marine mammals. DRC is a not-for-profit education and research facility.

Earthjustice Legal Defense Fund–Florida

www.Earthjustice.org/regional/tallahassee
111 S. Martin Luther King Jr. Blvd.
Tallahassee, FL 32301
ph: 850-681-0031
fax: 850-681-0020
eajusfl@earthjustice.org
Works to protect the places, natural resources, and wildlife of the planet and to defend the right of all people to a healthy environment by enforcing and strengthening environmental laws. This office is involved in a number of marine-related cases, including an effort to reduce accidental sea-turtle killings (called "bycatch") by long-line commercial fishing vessels.

Environmental Defense–Caribbean and East Florida Office

www.environmentaldefense.org
14630 SW 144 Terrace
Miami, FL 33186
Focuses on three themes fundamental to coastal conservation in the region: fisheries management and marine parks, coastal land use and habitat protection, and capacity building with local partners. Though most of the work is concentrated in Cuba and southeast Florida, significant projects are also underway in Mexico and Puerto Rico.

Everglades National Park

www.nps.gov/ever
40001 State Rd. 9336
Homestead, FL 33034-6733
ph: 305-242-7700
fax: 305-242-7711
EVER_Information@nps.gov
Features the only subtropical preserve in North America. The park spans the southern tip of the Florida peninsula and most of Florida Bay.

Florida Aquarium
www.flaquarium.org
701 Channelside Dr.
Tampa, FL 33602
ph: 813-273-4000
bclayton@flaquarium.org
Allows people of all ages and background to engage in
experiences that inspire a sense of wonder, understanding,
and stewardship of the aquatic environments.

Florida Chapter of the Wildlife Society
http://fltws.org/
551 N. Military Trail
West Palm Beach, FL 33415
vanderhoof@fltws.org
Works toward sustainable management of wildlife
resources and their habitats in Florida. Founded in 1968,
the FCTWS counts over 200 wildlife professionals as
members.

Florida Defenders of the Environment
www.fladefenders.org
4424 NW 13th, Ste. C-8
Gainesville, FL 32609-1885
ph: 352-378-8465
nick@fladefenders.org
Works to improve the quality of decision making affecting
Florida's environment. Formed in 1969, this coalition of volunteer scientists, economists, and lawyers has a history of
protecting the state's natural resources.

Florida Fishermen's Federation
www.floridafishermensfederation.com/
PO Box 1026
Panacea, FL 32346
Raypringle@earthlink.net
Represents the Mom and Pop fishing families in Florida and
fights for their sustainable way of life.

Florida Institute of Oceanography
830 First St. S
St. Petersburg, FL 33701
ph: 727-553-1100
fax: 727-553-1109
jogden@seas.marine.usf.edu
Supports and enhances Florida's coastal marine science
programs and keeps marine conservation at the forefront of
state discussions.

Florida League of Conservation Voters
www.floridalcv.org
PO Box 301
Tallahassee, FL 32302
ph: 850-222-0592
floridalcv@comcast.net
Serves as the nonpartisan advocacy and educational arm of
the environmental/conservation movement in Florida.

Florida Ocean Alliance
www.floridaoceanalliance.org
111 E. Las Olas Boulevard
Askew Tower, Ste. 709
Fort Lauderdale, FL 33301-2206
ph: 954-762-5268
fax: 954-762-5666
lalpert@fav.edu
Brings together government, academic, and private sectors
in Florida to protect and conserve coastal and ocean
resources for continued social and economic benefit.

Florida Oceanographic Society
www.floridaoceanographic.org
890 NE Ocean Blvd.
Stuart, FL 34996
ph: 772-225-0505
info@floridaoceanographic.org
Protects, preserves, and restores Florida's ocean and coastal
ecosystems through education, research, and personal
stewardship.

Florida PIRG
www.floridapirg.org
704 West Madison St.
Tallahassee, FL 32301
ph: 850-224-3321
fax: 850-224-1310
info@floridapirg.org
Educates citizens to participate in the political process.
Active in fisheries reform and other marine-related issues.

Florida Wildlife Federation
www.fwfonline.org
PO Box 6870
Tallahassee, FL 32314
ph: 850-656-7113
fax: 850-942-4431
wildfed@aol.com
Works to preserve, manage, and improve Florida's fish,
wildlife, soil, water, and plant life. This conservation education organization is affiliated with the National Wildlife
Federation.

Friends of FenHolloway River

http://groups.yahoo.com/group/hopeforcleanwater/
12677 Josh Ezell Rd.
Perry, FL 32348
ph: 850-584-7087
hope@gtcom.net
Fights to protect and restore one of the last industrial rivers
in the south.

Friends of Rookery Bay

www.rookerybay.org
300 Tower Rd.
Naples, FL 34113
ph: 239-417-6310
fax: 239-417-6315
info@rookerybay.org
Provides a basis for informed coastal decisions through
land management, restoration, research, and education. It
is a nonprofit citizens organization formed in 1987 to enlist
support for the Rookery Bay National Estuarine Research
Reserve.

Friends of the Everglades

www.everglades.org
7800 Red Rd., Ste. 215 K
Miami, FL 33143
ph: 305-669-0858
fax: 305-669-4108
info@everglades.org
Works to protect and restore the Everglades. Founded in
1969 by conservation pioneer and author Marjory
Stoneman Douglas, this volunteer river-of-grassroots organization has some 6,000 members.

Gulf Coast Conservancy

www.gulfcoastconservancy.org/Main.htm
PO Box 738
Aripeka, FL 34679
ph: 727-861-2621
cgula@tampabay.rr.com
Preserves the natural environment of coastal west-central
Florida.

Gulf Coast Environmental Defense

9 Gregory St.
Pensacola, FL 32501
Works to protect and preserve the natural resources of the
Florida Panhandle, especially its waters. This is a grassroots
environmental organization.

Guy Harvey Research Institute

www.nova.edu/ocean/ghri/
8000 N. Ocean Dr.
Dana Beach, FL 33004
ph: 954-262-3600
ghri@mail.ocean.nova.edu
Conducts high-quality, solution-oriented basic and applied
scientific research needed for effective conservation, biodiversity maintenance, restoration, and understanding of the
world's wild fishes.

Harbor Branch Oceanographic Institution

www.hboi.edu/index_02.html
5600 US 1 North
Ft. Pierce, FL 34946
ph: 772-465-2400
fax: 772-467-2061
webmaster@hboi.edu
Seeks to integrate the science and technology of the sea
with the needs of humankind, including discovery of new
medicines based on living marine organisms, without
depleting those resources.

Human Dolphin Institute

www.unc.edu/~segedy/hjsegedy/dolphin/index2.htm
118 Treasure Palms Dr.
Panama City, FL 32408
ph: 850-872-8003
fax: 850-872-8003
info@human-dolphin.org
Works for the protection and study of the local bottlenose
dolphin population.

Ichthyology at the Florida Museum of Natural History

www.flmnh.ufl.edu/fish/default.htm
Florida Museum of Natural History
Dickinson Hall, Ichthyology Dept. UFL
Gainesville, FL 32611
ph: 352-392-1721
fax: 352-846-0287
gburgess@flmnh.ufl.edu
Studies fish and their habitats at the museum of natural
history.

Indian Riverkeeper

www.indianriverkeeper.org
PO Box 1812
Jenson Beach, FL 34958
ph: 772-336-7284
keeper@indianriverkeeper.org
Works to protect and restore the waters of North America's
most diverse estuary, the Indian River Lagoon, and its tributaries, fisheries, and habitats through advocacy, enforcement, and citizen action.

International Game Fish Association
www.igfa.org
300 Gulf Stream Way
Dania, FL 33004
ph: 954-927-2628
fax: 954-927-4299
hq@igfa.org
Encourages youngsters to sport fish and maintains huge
databases on the subject. The association is actively
involved in various marine fish conservation efforts.

Izaak Walton League of America–Florida Division
http://www.iwla.org/
PO Box 97
Estero, FL 33928
ph: 239-992-1478
charlesdauray@mindspring.org
Works to protect both marine and terrestrial habitat for
future generations.

Key Deer Protection Alliance
www.keydeer.org
PO Box 430224
Big Pine Key, FL 33043
ph: 305-872-8888
fax: 305-872-8820
micalnnk@aol.com
Strives to protect the endangered Key deer.

League of Environmental Educators in Florida
http://leeflet.brinkster.net/
PO Box 540605
Lakeworth, FL 33454-0605
peggybirdhill@aol.com
Promotes environmental education in Florida at all levels
and through a variety of methods and resources.

Manatee Observation and Education Center
www.manateecenter.com
480 North Indian River Dr.
Fort Pierce, FL 34950
ph: 772-466-1600 x3333
manatee@manateecenter.com
A waterfront wildlife observation and nature education center located on Florida's east coast.

Mangrove Replenishment Initiative
www.mangrove.org
PO Box 510312
Melbourne Beach, FL 32951
ph: 321-431-6595
riley@mangrove.org
Develops self-sustaining, mangrove-stabilized shorelines
through active restoration efforts.

Marine Animal Rescue Society
www.marineanimalrescue.org
PO Box 833356
Miami, FL 33283
ph: 305-546-1111
mars@marineanimalrescue.org
Works for conservation of marine animals through rescue,
rehabilitation, research, and education.

Marine Discovery Center
www.marinediscoverycenter.org
162 North Causeway
New Smyrna Beach, FL 32169
ph: 866-257-4828
info@marinediscoverycenter.org
Conducts educational programs including eco-tours by pon-
toon boat, taking students and interested citizens through
the Indian River Lagoon.

The Marine Resources Council of East Florida
www.mrcirl.org/
3275 Dixie Hwy NE
Palm Bay, FL 32905
ph: 321-725-7775
fax: 321-725-3554
council@mrcirl.org
Offers several environmental programs and a specialized
library that is open to the public. Among its programs are
the Indian River Lagoonwatch, Shore Restoration,
Greenway Program, and Boaters for a Healthy Lagoon.

Miccosukee Tribe
www.miccosukeeresort.com
Miccosukee Information Center
Indian Village, Mile Marker 70
U.S. 41 Tamiami Trail
Miami, FL 33144
ph: 305-223-8380
Tribe@miccosukeeresort.com
Works to restore the Everglades and ensure that the high-
est standards of water quality are maintained as part of the
restoration effort. The Miccosukee Tribe of Seminole Nation
Indians has won a number of legal battles to protect the
Everglades and Florida Bay.

The Nature Conservancy–Florida
http://nature.org/wherewework/northamerica/states/florida/
222 S. Westmonte Dr., Ste. 300
Altamont Springs, FL 32714
ph: 407-682-3664
fax: 407-682-3077
mmelchiori@tnc.org
Works to protect coastal and marine resources. The Nature
Conservancy is a national organization headquartered in
Arlington, Virginia. Through land purchases, establishing

conservation easements and other means, it seeks to preserve the plants, animals, and natural communities that
represent the diversity of life on earth.

Newfound Harbor Marine Institute/Seacamp Association Inc.

www.seacamp.org
1300 Big Pine Ave.
Big Pine Key, FL 33043
ph: 305-872-2331
Offers marine science education and summer camp experiences to students ages twelve to seventeen. Founded in
1966, Seacamp is a nonprofit, marine science education
facility located on Big Pine Key in the tropical Florida Keys.

North Beach Neighborhood Association Inc.

201 Riverview Place
New Smyrna Beach, FL 32169
Works for the protection of the environment along
the coast.

The Ocean Conservancy–SE Atlantic Regional Office

www.oceanconservancy.org
449 Central Ave., Ste. 200
St. Petersburg, FL 33701
ph: 727-895-2188
dwhite@oceanconservancy.org

1000 Friends of Florida

www.1000fof.org
926 East Park Ave.
PO Box 5948
Tallahassee, FL 32314-5948
ph: 850-222-6277
fax: 850-222-1117
friends@1000fof.org
Maintains an experienced, professional staff to monitor
ongoing local, regional, and state growth management
activities that impact coastal Florida.

O.N.E.W.I.L.D.W.O.R.L.D.

www.Onewildworld.org
ph: 772-336-9252
onewildworld@aol.com
Works to "save paradise" and encourage environmental
stewardship through community events, beach and coastal
cleanups along the Treasure Coast, and a marine Web site
links directory.

Pelican Harbor Seabird Station

www.pelicanharbor.org
1279 NE 79th St. Causeway
Miami, FL 33138
ph: 305-751-9840
fax: 305-759-4461
wendy@pelicanharbor.org
Works to rescue, rehabilitate, and release sick and injured
seabirds.

Pensacola Gulf Coastkeeper

www.coastkeepers.org
811 W. Garden St.
Pensacola, FL 32501
ph: 850-429-8422
gckeeperrivers@cs.com
Protects the waterways of Northwest Gulf Coast Florida
through education, advocacy, and litigation.

Pew Institute for Ocean Science

www.pewocean.org
University of Miami
4600 Richenbacker Causeway
Miami, FL 33149
ph: 305-421-4163
fax: 305-421-4077
Promotes world-class scientific activity aimed at protecting
the world's oceans and the species that inhabit them.

Rare Species Conservatory Foundation

www.rarespecies.org
PO Box 1371
Loxahatchee, FL 33470
ph: 561-790-5864
fax: 561-792-2122
info@rarespecies.org
Works to preserve marine and terrestrial biodiversity
through grassroots conservation programs rooted in
sound science.

Reef Environmental Education Foundation (REEF)
www.reef.org
PO Box 246
Key Largo, FL 33037
ph: 305-852-0030
fax: 305-852-0301
lad@reef.org
Strives to preserve the marine environment. The recreational divers who belong to this grassroots, nonprofit organization regularly conduct fish biodiversity and abundance surveys during their dives.

Reefguardian International
www.reefguardian.org
ph: 305-358-4600
fax: 305-358-4600
info@ReefGuardian.org
Works to protect coral reefs and their marine life.

Reef Relief
www.reefrelief.org
PO Box 430
Key West, FL 33041
ph: 305-294-3100
fax: 305-293-9515
info@reefrelief.org
Works to preserving and protecting living coral reef ecosystems through local, regional, and global efforts. This nonprofit membership organization focuses on rigorous science to educate the public and advocates to policymakers to achieve conservation, protection, and restoration of coral reefs.

Sanibel-Captiva Conservation Foundation
www.sccf.org
3333 Sanibel-Captiva Rd.
PO Box 839
Sanibel, FL 33957
ph: 239-472-2329
fax: 239-472-6241
sccf@sccf.org
Seeks to preserve natural resources and wildlife habitat on and around the barrier islands.

Save a Turtle
www.save-a-turtle.org
PO Box 361
Islamorada, FL 33036
ph: 305-743-6056
newsletter@save-a-turtle.org
Aims to preserve sea turtles and their habitat around the Florida Keys through education, beach cleanup, turtle nesting surveys, stranding response, and other efforts.

Save Our Seabirds
www.seabirdrehab.org
840 3rd Ave. S
Tierra Verde, FL 33715
ph: 727-864-0679
sosinc@tampabay.rr.com
Works with injured and disabled seabirds to rectify their misadventures with mankind. SOS (formerly the Pinellas Seabird Rehabilitation Center) was founded in May 1990. The group works with power companies, petroleum products manufacturers, and transporters to educate them about a wildlife responder's needs prior to an environmental incident. SOS also trains volunteers for oiled wildlife response in the volunteers' home geographic areas.

Save the Manatee Club
www.savethemanatee.org
500 N. Maitland Ave., Ste. 210
Maitland, FL 32751
ph: 800-432-5646
fax: 407-539-0990
education@savethemanatee.org
Promotes public awareness and education about endangered manatees; manatee research, rescue, and rehabilitation; and advocacy and legal action to ensure better protection and habitat for manatees. SMC is a 40,000 member nonprofit organization established in 1981 by former Governor Bob Graham and singer/songwriter Jimmy Buffett.

Seakeepers Society
www.seakeepers.org
1800 S.E. 10th Ave., Ste. 213
Fort Lauderdale, FL 33316
ph: 954-766-7100 x101
fax: 954-252-4595
info@seakeepers.org
Protects the oceans by equipping ships and yachts with ocean and atmospheric monitoring sensors.

The Seaside Institute
www.theseasideinstitute.org
PO Box 4730 Seaside Branch
Santa Rosa Beach, FL 32459
ph: 850-231-2421
fax: 850-231-1884
lwalker@theseasideinstitute.org
Promotes building sustainable communities in cities and towns through design, education, and the arts. Seaside itself is seen as a model of sustainable development along the shore.

Seaturtle Preservation Society
www.seaturtlespacecoast.org
PO Box 510988-0988
Melbourne Beach, FL 32951-0988
ph: 321-676-1701
office@seaturtlespacecoast.org
Educates the public about marine turtles through lectures, exhibits, and turtle watches during nesting season.

SeaWorld of Florida
www.buschgardens.com/seaworld/fla
7007 SeaWorld Dr.
Orlando, FL 32821
ph: 407-351-3600
Promotes marine conservation through entertainment.

Sierra Club National–Florida
www.sierraclub.org
475 Central Ave., Ste. M-1
St. Petersburg, FL 33701
ph: 727-824-8813
fax: 727-824-0936
se.field@sierraclub.org
Focuses on "Florida's Water Heritage," particularly on the Gulf and Atlantic coasts and the Keys, as well as drinking water, rivers, and lakes. Its chapters are involved in most major marine issues from coastal development to coral protection.

Sierra Club National–South Florida/Everglades
www.sierraclub.org
2700 SW 3rd Ave., Ste. 2F
Miami, FL 33129
ph: 305-860-9888
fax: 305-860-9862
se.field@sierraclub.org

Sierra Club–Florida, Big Bend
http://florida.sierraclub.org/bigbend/
PO Box 15732
Tallahassee, FL 32317
ph: 850-297-2052

Sierra Club–Florida, Broward Group
http://.florida.sierraclub.org/broward
PO Box 350432
Ft. Lauderdale, FL 33335
ph: 954-858-0160

Sierra Club–Florida, Calusa Group
http://florida.sierraclub.org/calusa
27277 Arroyal Rd.
Bonita Springs, FL 34135
ph: 239-992-1565
bleegruninger@comcast.net

Sierra Club–Florida, Loxahotchee Group
http://florida.sierraclub.org/loxahotchee
ph: 561-833-0405
johnkay@mindspring.com

Sierra Club–Florida, Manatee/Sarasota Group
http://florida.sierraclub.org/sarasota
PO Box 3485
Sarasota, FL 34236-3485
ph: 941-925-9000
vmplbk@aol.com

Sierra Club–Florida, Miami Group
http://florida.sierraclub.org/miami
PO Box 43-0741
South Miami, FL 33243-0741
ph: 305-667-7731

Sierra Club–Florida, Nassau County Group
http://florida.sierraclub.org/nassaucounty
PO Box 38
Ferndina Beach, FL 32035
ph: 904-277-4187
titcombe@bellsouth.net

Sierra Club–Florida, Nature Coast
www.florida.sierraclub.org/naturecoast
482 Homosassa Springs, FL 34447
ph: 352-628-7540

Sierra Club–Florida, Northeast
http://florida.sierraclub.org/northeast
2029 3rd St. N
Jacksonville Beach, FL 32250
ph: 904-399-1520
neflsierraclub@yahoo.com

Sierra Club–Florida, Northwest Group
http://florida.sierraclub.org/northwest
PO Box 4907
Seaside, FL 32459

Sierra Club–Florida, Suncoast
http://florida.sierraclub.org/suncoast
PO Box 4183
St. Petersburg, FL 33731
ph: 727-945-0999
cathy_bam@hotmail.com

Sierra Club–Florida, Tampa Bay Group

http://florida.sierraclub.org/tampabay
PO Box 1948
Tampa Bay, FL 33601
ph: 813-253-3555
delcope@tampabay.rr.com

Sierra Club–Florida, Turtle Coast Group

http://florida.sierraclub.org/turtlecoast
PO Box 061887
Palm Bay, FL 32906
ph: 321-631-3916
conservation@turtlecoast.org

Smithsonian Marine Station

www.sms.si.edu
701 Seaway Dr.
Fort Pierce, FL 34949
ph: 772-465-6630
fax: 772-461-8154
paul@sms.si.edu
Supports and conducts scholarly research in the marine sciences, including collection, documentation, and preservation of south Florida's marine biodiversity and ecosystems, as well as education, training, and public service.

Society for Marine Mammalogy

www.marinemammalogy.org
admin@marinemammalogy.org
Promotes educational, scientific, and managerial advancement of marine mammal science.

South Florida Resource Conservation Council

www.sfrcd.org
15600 SW 288th St., Ste. 402
Homestead, FL 33033
ph: 305-246-4319
fax: 305-245-2473
ggarvey@sfrcd.org
Improves community quality of life through resource conservation and development projects including land conservation, water management, community development, and environmental enhancement.

St. Johns Riverkeeper

www.stjohnsriverkeeper.org
2800 University Boulevard North
Jacksonville, FL 32211
ph: 904-256-7591
narming@ju.edu
Seeks to protect, preserve, and restore the ecological integrity of the St. Johns River watershed for current users and future generations through advocacy and citizen action.

Surfrider Foundation–Florida, First Coast

PO Box 51225
Jacksonville Beach, FL 32240-1225
ph: 904-343-8325
scott@jaxsurfrider.org
Uses conservation, activism, research, and education to encourage people to protect and enjoy the world's waves and beaches. The Surfrider Foundation is a nonprofit environmental organization whose membership consists mainly of surfers. Its scope has expanded to include protecting wetlands, bird life, and beaches as well as surf spots. It has sixty-six chapters worldwide, including six chapters in Florida.

Surfrider Foundation–Florida, Orlando Chapter

152 Stone Gable Circle
Winter Springs, FL 32708
ph: 407-718-5952

Surfrider Foundation–Florida, Palm Beach County Chapter

www.surfriderpbc.org
PO Box 33687
Palm Beach Gardens, FL 33420-3687
ph: 561-889-6196

Surfrider Foundation–Florida, Sebastian Chapter

www.surfrider.org/sebastianinlet
PO Box 372067
Satellite Beach, FL 32937
ph: 321-868-7897
ggordon3@cfl.rr.com

Surfrider Foundation–Florida, South Florida Chapter

http://storm.rsmas.miami.edu/~cook/Surfrider
58 NE 92nd St.
Miami Shores, FL 33138
ph: 305-672-3305
surfridermiami@yahoo.com

Tampa Bay Watch Inc.

www.tampabaywatch.org
3000 Pinellas Bayway S
Tierra Verde, FL 33715
ph: 727-867-8166
fax: 727-867-8188
info@tampabaywatch.org
Works to protect and restore the marine environments of the Tampa Bay Estuary by expanding community involvement in hands-on restoration and protection projects around the bay, the largest open-water estuary in the state of Florida.

Tropical Audubon Society

www.tropicalaudubon.org
5530 Sunset Dr.
Miami, FL 33143
ph: 305-667-7337
Works to protect the natural world and promotes wise
stewardship of natural resources, especially native plants
and animals. This organization was established in 1947 as
a chapter of the Audubon Society of Florida.

Turtle Time Inc.

www.turtletime.org
PO Box 2621
Fort Myers, FL 33932
eve@turtletime.org
Strives for the continued survival of the Loggerhead
sea turtle.

Wild Dolphin Project

www.wilddolphinproject.org
PO Box 8436
Jupiter, FL 33468
ph: 561-575-5660
wilddolphinproject@earthlink.net
Studies a specific pod of Atlantic bottlenose dolphins.

WildLaw–Florida Office

www.wildlaw.org
1415 Devils Dip
Tallahassee, FL 32308
ph: 850-878-6895
fax: 850-878-6895
wildlawfl@comcast.net
Reviews and exposes the violations of permitted industry
and government entities in southern states and ensures
that environmental laws are adhered to.

Wildlife Foundation of Florida

www.wildlifefoundationofflorida.com
PO Box 11010
Tallahassee, FL 32302
ph: 850-922-1066
foundation@myfwc.com
Provides funding and promotional support for the Florida
Fish and Wildlife Conservation Commission.

World Wildlife Fund–South Florida

http://www.wwfus.org/wildplaces/sfla/projects.cfm
8075 Overseas Highway
Marathon, FL 33050
Directs efforts to saving the Everglades and the Florida
Keys. WWF's South Florida program uses an innovative conservation vision that brings together communities for the
benefit of people and the environment.

GEORGIA

Altamaha Riverkeeper, Inc.

www.Altamahariverkeeper.org
PO Box 2642
Darien, GA 31305
ph: 912-437-8164
stewards@altamahariverkeeper.org
Works to protect, defend, and restore Georgia's biggest
river.

Center for a Sustainable Coast

www.sustainablecoast.org
221B Mallory St.
Saint Simons Island, GA 31522
ph: 912-638-3612
susdev@gate.net
Seeks to improve the responsible use, protection, and conservation of coastal Georgia's resources—natural, historic,
and economic.

Coastal Conservation Association–Georgia

www.ccaga.org
515 Denmark St., Ste. 300
Statesboro, GA 30458
ph: 912-764-6222
fax: 912-764-6497
info@ccaga.org
Promotes the preservation, conservation, restoration, and
protection of the marine fisheries and habitats of the
Georgia coast.

Coastal Environmental Organization of Georgia

www.theceo.org
231 Georgia Hwy. 204
Pembroke, GA 31321
ph: 912-858-2356
info@theceo.org
Strives to alleviate the growing threats to irreplaceable
coastal resources by educating citizens and policy makers
around issues like the Savannah Harbor expansion project
and aquifer storage and recovery by private water
companies.

Coastal Georgia Center for Sustainable Development

PO Box 598
Darien, GA 31305
Seeks ways to maintain the coastal region of Georgia in a
natural state while also protecting the economic health
and well-being of local communities that depend on its
abundance.

Coastal Georgia Land Trust Inc.

www.galandtrust.org
428 Bull St., Ste. 210
Savannah, GA 31405
ph: 912-231-0507
Works with landowners to help protect their lands and waters. Now known as the Georgia Land Trust.

Georgia Conservancy

www.gaconservancy.org/Home/Home.asp
817 West Peachtree St., Ste. 200
Atlanta, GA 30308
ph: 404-876-2900
fax: 404-872-9229
mail@gaconservancy.org
Ensures that Georgians have healthy air, clean water, unspoiled wild places, and community green space and works on alternatives to rapid coastal development. This is a statewide environmental organization.

Georgia Environmental Organization

www.gaenv.org
3185 Center St.
Smyrna, GA 30080-7039
ph: 404-605-0000
fax: 404-350-9997
info@gaenv.org
Seeks to preserve and protect Georgia's environment through education, collaboration, research, planning, legislation, and grassroots organizing.

Georgia River Network

www.garivers.org
126 S. Milledge Ave., Ste. E3
Athens, GA 30605
ph: 706-549-4508
fax: 706-549-7791
info@garivers.org
Works to protect and restore rivers and watersheds by building local group capacity and providing statewide water policy analysis.

Georgia PIRG

www.georgiapirg.org
1447 Peachtree St. NE, Ste. 304
Atlanta, GA 30309
ph: 404-892-3573
fax: 404-892-5201
info@georgiapirg.org
Works for the public interest through a range of citizen action and mobilization on a range of issues including protection of coastal communities.

Gray's Reef National Marine Sanctuary

www.graysreef.nos.noaa.gov
10 Ocean Science Circle
Savannah, GA 31411
ph: 912-598-2345
fax: 912-598-2367
graysreef@noaa.gov
Features one of the largest near-shore live-bottom reefs of the southeastern United States.

National Wildlife Federation–Southeastern Natural Resource Center

www.nwf.org
1330 W. Peachtree St. NE, Ste. 475
Atlanta, GA 30309
ph: 404-876-8733
fax: 404-892-1744
Works with NWF affiliates and other organizations to conserve natural public resources including coastal resources, for future generations.

Reef Ball Foundation

www.Reefball.org
603 River Overlook Rd.
Woodstock, GA 30188
kathy@reefball.org
Seeks to restore the world's ocean ecosystems and protect natural reef systems through preservation, reef ball technology (providing concrete structures for coral growth), innovative public education, and community involvement. It has conducted over 3,500 reef restoration projects worldwide from Florida to Malaysia. Some of the artificial reef balls mix human ashes with the concrete to create "living memorials."

Savannah Riverkeeper

www.savannahriverkeeper.org
1226 River Ridge Rd.
Augusta, GA 30909
ph: 706-364-5253
frank.carl@savannahriverkeeper.org
Works to preserve, protect, and restore the Savannah River.

Save the Beach

65 Village Green St.
Simons Island, GA 31522
ph: 912-638-1408
fax: 912-638-1408
Works to protect the beach and other natural resources that make this area so attractive.

Sierra Club–Georgia, Coastal Group

http://georgia.sierraclub.org/coastal
Coastal Group Sierra Club
3 Belleview Ct.
Savannah, GA 31406
ph: 912-351-7436
johnrhs@aol.com
Works to protect and enjoy Georgia's unique salt marshes and other coastal resources.

Southern Environmental Law Center

www.southernenvironment.org
The Chandler Building
127 Peachtree St., Ste. 605
Atlanta, GA 30303
ph: 404-521-9900
fax: 404-521-9909
Strengthens environmental laws and policies throughout the South. The center works with partners to protect coastal marshes from projects that threaten the area's water quality and wildlife.

Upper Chattahoochee Riverkeeper

www.ucriverkeeper.org
916 Joseph Lowery Blvd.
3 Puritan Mill, Ste. 3
Atlanta, GA 30318
ph: 404-352-9828
fax: 404-352-8676
Uses advocacy, education, research, communication, cooperation, monitoring, and legal actions to protect and preserve the Chattahoochee River and its watersheds from its headwaters to the sea.

HAWAII

Community Conservation Network

http://conservationpractice.org/
PO Box 4674
Honolulu, HI 96812
ph: 808-528-3700
fax: 808-528-3701
info@conservationpractice.org
Helps local communities and their partners to sustain vital ecosystems and resources, including remote atoll ecosystems and marine fishing community resources, by fostering relationships and building capacity that results in improved long-term conservation, management effectiveness, and human security.

Dolphin Institute

www.dolphin-institute.org
420 Ward Ave., Ste. 212
Honolulu, HI 96814
ph: 808-593-2211
fax: 808-593-2211
correspondence@dolphininstitute.com
Dedicated to the preservation of dolphins, whales, and other marine mammals, and to the education of people whose attitudes and activities affect the well-being of the animals. People gather here to work and learn from the dolphins.

Earthjustice Hawaii Office

www.earthjustice.org/regional/honolulu/
223 South King St., #400
Honolulu, HI 96813
ph: 808-599-2436
eajushi@earthjustice.org
Seeks to protect the magnificent places, natural resources, and wildlife and to defending the right of all people to a healthy environment. Earthjustice is a nonprofit public interest law firm. This office plays a vital role in assuring the long-term protection of the Northwest Hawaiian Islands Ecosystem Reserve.

EarthTrust

www.earthtrust.org
1118 Maunawili Rd.
Kailua, HI 96734
ph: 808-261-5339
fax: 206-202-3893
earthtrustinformation@yahoo.com
Works as a global innovator for wildlife and the environment. Earth Trust was the first environmental organization to document the high-seas driftnet fishery that was destroying the north Pacific ecosystem. Their work helped lead to a UN ban on the destructive 40-mile-long nets.

Environmental Defense–Hawaii

www.environmentaldefense.org/Hawaii
PO Box 520
Waimanalo, HI 96795
ph: 808-262-7128
stephfeenvironmentaldefense.org
Focuses on helping establish and strengthen community networks to protect marine ecosystems and cultural resources in the Asia and Pacific Islands region, including areas such as the Northwest Hawaiian Islands Ecosystem Reserve.

Environment Hawaii

www.environment-hawaii.org
72 Kapi'olani St.
Hilo, HI 96720
ph: 808-934-0115
Provides environmental news for Hawaii including extensive coverage of marine issues.

Friends of He'eia State Park

www.friendsofheeia.com
PO Box 698
Kaneohe, HI 96744
ph: 808-247-3156
fax: 808-247-8510
admin@friendsofheeia.com
Fosters a sense of stewardship toward the natural and cultural resources of the land and sea. This group is a non-profit organization.

Hawaii Audubon Society

www.hawaiiaudubon.com
850 Richards St., Ste. 505
Honolulu, HI 96813-4709
ph: 808-528-1432
fax: 808-537-5294
hiaudsoc@pixi.com
Works to protect habitat for the largest number of endangered species of any state in the nation. Audubon is a national network of community-based nature centers and chapters, scientific and educational programs, and advocacy on behalf of areas sustaining important bird populations.

Hawaii Conservation Association

www.konatournaments.com
74-425 Kealakehe Pkwy., Ste. 15
Kailua Kona, HI 96740
ph: 808-960-0978
jody@konatournaments.com
Helps instill an inherent commonsense conservation ethic toward Hawaii's ocean resources.

Hawaii Maritime Center

www.bishopmuseum.org
Pier 7 Honolulu Harbor
Honolulu, HI 96813
ph: 808-523-6151
fax: 808-536-1519
Features the story of Polynesian seafaring, Hawaiian water culture, whaling, shipping, and surfing.

Hawaii's Thousand Friends

305 Hahani St., #282
Kailua, HI 96734
ph: 808-262-0682
fax: 808-262-0682
htf@lava.net
Ensures that land-use decisions are made to protect the environment, cultural, and natural resources, including beaches, bays, and coral reefs.

Hawaii Wildlife Fund

www.wildhawaii.org
PO Box 637
Paia, Maui, HI 96779
wild@aloha.net
Works to preserve Hawaii's native wildlife through research, conservation, and education.

International Marinelife Alliance

www.marine.org/
Stangenwald Bldg., Ste. 610
Honolulu, HI 96813
info@marine.org
Works to end human influences that degrade and destroy the world's marine environment while improving the overall quality of life for people and communities living in and dependent on coastal and marine resources.

KAHEA: The Hawaiian-Environmental Alliance

www.kahea.org
PO Box 27112
Honolulu, HI 96827
ph: 808-524-8220
fax: 808-524-8221
kahea-alliance@hawaii.rr.com
Seeks to protect customary and traditional rights and the fragile island and marine environment. KAHEA is an alliance of Kanaka Maoli (Native Hawaiian) cultural practitioners and environmental activists.

Kai Makana

www.kaimakana.org
PO Box 22719
Honolulu, HI 96823
ph: 808-261-8939
info@kaimakana.org
Educates and mobilizes the public to better understand and preserve marine life.

Kewalo Basin Marine Mammal Lab

www.dolphin-institute.org
94-117 Kekai Place
Kapolei, HI 96707
ph: 808-679-3690
fax: 808-679-3690
correspondence@dolphin-institute.org
Studies dolphin behavioral capacity. KBMML pioneered the study of humpback whales off Hawaii in 1975. It established the Dolphin Institute as a not-for-profit corporation in 1993.

Kohanaiki 'Ohana
www.kohanaiki.org
PO Box 4753
Kailua-Kona, HI 96745
ph: 808-325-0844
fax: 808-325-6322
kohanaiki@kona.net
Perpetuates stewardship of natural resources, human rights,
and environmental justice. Also sponsors beach cleanups
and coastal trail restoration.

Marine Aquarium Council
www.aquariumcouncil.org
923 Nu'uanu Ave.
Honolulu, HI 96817
ph: 808-550-8217
fax: 808-550-8317
info@aquariumcouncil.org
Works to conserve coral reefs and other marine ecosystems
by creating standards and certification for those engaged in
collection and care of ornamental marine life from reef to
aquarium.

Maui Coastal Land Trust
RR 1 Box 398
Kula, HI 96790
ph: 808-871-6852
starr@maui.net
Works today to preserve Maui's open space and coastal
resources forever.

Na Pali Coast Ohana
www.napali.org
PO Box 452
Lihue, HI 96766
ph: 808-241-7245
info@napali.org
Strives to preserve the natural and cultural resources of Na
Pali Coast State Park, a stunning wilderness area on the
northwest coast of Kauai, Hawaii.

**Native Hawaiians for the Preservation of Hawaiian
Ecosystems ('Ahahui Malama I Ka Lokahi)**
http://alohahosting.net/~ahahui/
PO Box 61578
Honolulu, HI 96839-1578
aml@aecos.com
Works to develop, promote, and practice a native Hawaiian
conservation ethic relevant to today that is responsible to
both Hawaiian culture and science. This ethic protects
native cultural and natural heritage and is expressed
through research, education, and active stewardship.

The Nature Conservancy–Hawaii
http://nature.org/wherewework/northamerica/states/hawaii/
923 Nu'uanu Ave.
Honolulu, HI 96817
ph: 808-537-4508
fax: 808-537-2019
hawaii@tnc.org
Works to preserve plants, animals, and natural communities
in the state with the most endangered species of native
plants and animals in the nation. The Nature Conservancy
is a national organization headquartered in Arlington,
Virginia.

Pacific Island Land Institute
www.pilipacific.org
270 Ku'ulei Rd., #201
Kailua, HI 96734
info@pilipacific.org
Protects landowners and communities in Hawaii and other
Pacific Islands in their efforts to protect, preserve, and
restore the islands' natural, cultural, and agricultural
resources.

Pacific Whale Foundation
www.pacificwhale.org
300 Maalaea Rd., Ste. 211
Wailuku, HI 96793
ph: 808-249-8811
fax: 808-243-9021
info@pacificwhale.org
Works to save the planet's oceans and the life they contain.

Project SEA-Link
www.projectsealink.org
ph: 808-669-9062
info@projectsealink.org
Strives to promote marine science, education, and awareness by providing links between students, teachers, scientists, the local community, the general public, other nonprofit organizations, and governmental agencies.

Save Our Seas
www.saveourseas.org
PO Box 813
Hanalei, HI 96714
ph: 808-651-3452
sos@saveourseas.org
Uses education and research to preserve, protect, and
restore the world's oceans for future generations. SOS is an
international Hawaii-based nonprofit organization.

Sea Life Park Hawaii

www.sealifeparkhawaii.com
41–202 Kalanianaole Hwy., Ste. 7
Waimanalo, HI 96795
ph: 866-365-7446
info@sealifeparkhawaii.com
Offers educational opportunities to learn about marine
wildlife.

Sierra Club–Hawaii Chapter, Honolulu

www.hi.sierraclub.org
PO Box 2577
Honolulu, HI 96803
ph: 808-538-6616
mikulina@lava.net
Explores, enjoys, and protects the planet. "The Blue Water
Campaign," one of the chapter's major efforts, helps protect Hawaii's waters from runoff and pollution through
community outreach, education and involvement, cooperation with government agencies, and new legislation. The
Sierra Club is America's oldest, largest, and most influential
grassroots environmental group.

Surfrider Foundation–Hawaii, Maui

www.surfrider.org/maui
PO Box 790549
Paia, HI 96779
Uses conservation, activism, research, and education to
encourage people to protect and enjoy the world's waves
and beaches. The Surfrider Foundation is a nonprofit environmental organization whose membership consists mainly
of surfers. Its scope has expanded to include protecting
wetlands, bird life, and beaches as well as surf spots. It has
sixty-six chapters worldwide, including two chapters in
Hawaii.

Surfrider Foundation–Hawaii, Oahu

www.surfrider.org/oahu/
66-077 Wana Place
Haleiwa, HI 96712
ph: 808-637-4151

Trout Unlimited–Hawaii

www.tuhi.us/
268 Aikahi Place
Kailua, HI 96816
ph: 808-254-8701
fax: 808-254-9517
arkl.baker@verizon.net
Supports sustainable resource use in Hawaii and elsewhere.

Universal Cetacean Institute

www.uci-endingcaptivity.org
PO Box 1359
Kilauea, Kauai, HI 96754
ph: 808-828-0370
uci@compuserve.com
Educates the public about the truth regarding captive
marine mammal display industries.

Waikiki Aquarium

http://waquarium.otted.hawaii.edu/
2777 Kalakaua Ave.
Honolulu, HI 96815
ph: 808-923-9741
fax: 808-923-1771
Seeks to help people of all ages to understand, love, care
for, and work to protect the life of the ocean through a
commitment to excellence in educational and entertaining
experiences, research, and conservation.

West Hawaii Fisheries Council

PO Box 489
Kailua Kona, HI 96745
ph: 808-325-5000
fax: 808-325-7023
Manages fishery activities to ensure sustainability and
enhance nearshore resources, minimize resource depletion
and conflicts of use, provide for involvement of the community in decisions, and encourage scientific research and
monitoring of the nearshore environment. WHFC is a
community-based management group based on the Big
Island.

Whaleman Foundation

www.whaleman.org
PO Box 1670
Lahaina, HI 96767
ph: 808-661-8859
fax: 808-661-8859
whaleman@maui.net
Works to preserve and protect cetaceans (dolphins, whales,
and porpoises) and their critical habitats.

Whale Trust

www.whaletrust.org
300 Paani Place
Paia, HI 96799
ph: 808-873-9600
fax: 808-873-9603
info@whaletrust.org
Supports field research on whales and their environment.

The Wild Dolphin Foundation
87-1286 Farrington Hwy.
Waianae, HI 96792
ph: 808-668-4075
info@wilddolphin.org
Protects and restores the natural habitats of dolphins
through research, advocacy, and public education. Projects
include the Dolphin ID Web site, habitat and reef surveys,
and video documentation.

zero impact Productions
www.zeroimpactproductions.com/op.html
851 S Kihei Rd., #0115
Kihei, Hawaii 96753
marie@zeroimpactproductions.com
Works to inspire greater respect for the underwater world
through quality underwater video and documentary
production.

ILLINOIS
John G. Shedd Aquarium
www.sheddaquarium.org/
1200 S. Lake Shore Dr.
Chicago, IL 60605
ph: 312-939-2435
fax: 312-939-3793
contactus@sheddaquarium.org
Promotes the enjoyment, appreciation, and conservation of
aquatic life and environments through education, exhibits,
and research.

IOWA
National Mississippi River Museum and Aquarium
www.rivermuseum.com
350 E. 3rd St
Dubuque, IA 52001
ph: 563-557-9545
fax: 563-583-1241
info@rivermuseum.com
Teaches the story of the people, natural history, and leg-
ends of the Mississippi River from its headwaters to
the sea.

Science and Environmental Health Network
www.sehn.org
3704 W. Lincoln Way, #282
Ames, IA 50014
ph: 515-268-0600
fax: 515-268-0604
info@sehn.org
Seeks the wise application of science to the protection of
the environment and public health, including the reduction
of agricultural chemicals that pollute the Mississippi River
and Gulf of Mexico. A consortium of North American environmental organizations founded SEHN in 1994.

LOUISIANA
Atchafalaya Basinkeeper
32675 Gracie Ln., Unit D
Plaquemine, LA 70764
ph: 225-659-2499
basinkeeper@aol.com
Seeks to identify sources of pollution, and to confront and,
if necessary, sue polluters to clean up coastal Louisiana.
Atchafalaya Basinkeeper is part of the Waterkeeper Alliance.

Audubon Aquarium of the Americas
www.auduboninstitute.org/aoa
PO Box 4327
New Orleans, LA 70178
ph: 504-378-2694
air@auduboninstitute.org
Offers educational and conservation programs and
resources. The central tank is designed to resemble the legs
of an offshore oil rig.

Audubon Council of Louisiana
www.louisianaaudubon.org
355 Napolean St.
Baton Rouge, LA 70802
ph: 504-861-8465
dlafleur@lpb.org
Works to protect bottomland hardwoods, wetlands habitat,
and endangered species of Louisiana, and to alert the public to the dangers of mercury-contaminated fish. Organized
in 1989, the LAC comprises Audubon chapters in the state
and nonchapter members of the National Audubon Society.

Audubon Society–Orleans
www.jjaudubon.net
801 Rue Dauphne
Metaire, LA 70005
ph: 504-834-BIRD
fax: 504-834-BIRD
jacoulson@aol.com
Seeks to conserve and restore natural ecosystems, while
focusing on birds, other wildlife, and their habitats for the
benefit of humanity and the earth's biological diversity. This
group is a chapter of the National Audubon Society.

Coalition to Restore Coastal Louisiana
www.crcl.org
746 Main St., Ste. B101
Baton Rouge, LA 70802
ph: 225-344-6555
coalition@crcl.org
Works to restore sustainability to coastal Louisiana. The
coalition represents participants from many different interests among business, local government, scientists, and
concerned citizens from the conservation and religious
communities.

Family Fishermen (Association of)
4927 Deborah Ann Dr.
Barataria, LA 70036
familyfishermen@cox.net
Seeks to protect the resource through reduction of runoff pollution from the Mississippi River and to represent its members.

Gulf Restoration Network
www.healthygulf.org
338 Baronne St., Ste. 200
New Orleans, LA 70112
ph: 504-525-1528
fax: 504-525-0833
amy@healthygulf.org
Works to restore and protect the resources of the Gulf of Mexico. GRN is a diverse network of local, regional and national groups, with members in the five Gulf states of Texas, Louisiana, Mississippi, Alabama, and Florida.

Lake Pontchartrain Basin Foundation
www.saveourlake.org
PO Box 6965
Metairie, LA 70009
ph: 504-836-2215
fax: 504-836-7070
lpbfinfo@saveourlake.org
Strives to preserve the Lake Pontchartrain Basin.

Louisiana Bayoukeeper
PO Box 207
Barataria, LA 70036
ph: 504-689-7880
bayoukeeper@cox.net
Works to protect the famed but threatened bayous of southern Louisiana. This group is part of the Waterkeeper Alliance.

Louisiana Environmental Action Network (LEAN)/Lower Mississippi Riverkeeper
www.leanweb.org
PO Box 66323
Baton Rouge, LA 70896
ph: 225-928-1315
fax: 225-922-9247
lean@leanweb.org
Fosters cooperation between organizations to mend the environmental problems in Louisiana. Riverkeeper is a working project of LEAN.

Louisiana Wildlife Federation
www.lawildlifefed.org
PO Box 65239
Audubon Station
Baton Rouge, LA 70896-5239
ph: 225-344-6707
lwf@lawildlifefed.org
Works to conserve the natural resources of Louisiana primarily through education and advocacy, with particular emphasis on fish and wildlife and their habitats.

LUMCON–Louisiana Universities Marine Consortium
www.lumcon.edu
8124 Highway 56
Chauvin, LA 70344
ph: 504-851-2800
fax: 504-851-2874
information@lumcon.edu
Stimulates Louisiana's activities in marine research and education. Scientists at the consortium were the first to identify and describe the Gulf of Mexico's nutrient-fed dead zone.

Mississippi River Basin Alliance
www.mrba.org
PO Box 4268
New Orleans, LA 70178
ph: 504-588-9008
fax: 504-862-8455
A coalition of grassroots organizations working to bring people together to save the Mississippi River, from the headwaters of northern Minnesota to the Gulf of Mexico and beyond.

The Nature Conservancy–Louisiana
http://nature.org/wherewework/northamerica/states/louisiana/
PO Box 4125
Baton Rouge, LA 70821
ph: 225-338-1040
fax: 225-338-0103
lafo@tnc.org
Works to protect over 325,000 acres of critical natural lands, including Cypress islands. The Nature Conservancy is a national organization headquartered in Arlington, Virginia. Through land purchases, establishing conservation easements, and other means, it seeks to preserve the plants, animals, and natural communities that represent the diversity of life on earth.

North Baton Rouge Citizen's Association
421 Springfield Rd.
Baton Rouge, LA 70807
ph: 504-775-0341
fax: 504-775-0341
Works to protect natural resources along the Mississippi
River near the rapidly subsiding coastline.

North Shore Coast Watch
165 Swallow St.
Covington, LA 70433
Protects the north shore of Lake Pontchartrain.

Sierra Club–Delta Chapter
http://louisiana.sierraclub.org
PO Box 19469
New Orleans, LA 70179-0469
ph: 504-836-3062
chapter-info@louisianasierraclub.org
Explores, enjoys, and protects the planet and works for the
protection of Louisiana including its threatened coast
through education and advocacy.

Slidell Working Against Major Polluters
523 Legendre Dr.
Slidell, LA 70460
Works against polluters in Louisiana whose emissions
and runoff threaten both human health and the marine
environment.

Tulane Environmental Law Clinic
www.tulane.edu/~telc
6329 Freret St.
New Orleans, LA 70118
ph: 504-865-5789
fax: 504-862-8721
rdayries@law.tulane.edu
Advocates and litigates for protection of the health and
habitat of river and coastal communities. This is the largest
most active environmental law clinic in the country.

MAINE
Atlantic Salmon Federation
www.asf.ca
14 Maine St., Ste. 308
Brunswick, ME 04011
ph: 207-725-2833
fax: 207-725-2967
asfme@blazenetme.net
Promotes the conservation and wise management of wild
Atlantic salmon and their environment.

Audubon Society–Maine
www.maineaudubon.org
20 Gilsland Farm Rd.
Falmouth, ME 04105
ph: 207-781-2330
fax: 207-781-0974
info@maineaudubon.org
Works to conserve and restore natural ecosystems, focusing
on birds, other wildlife, and their habitats for the benefit of
humanity and the earth's biological diversity. Educational
focus includes the Hog Island camp and Teachers' Resource
Center.

Bigelow Laboratory for Ocean Sciences
www.bigelow.org
PO Box 475
West Boothbay Harbor, ME 04575-0475
ph: 207-633-9600
fax: 207-633-9641
webmaster@bigelow.org
Promotes a comprehensive approach to oceanography that
stresses the interdependence of marine biology, chemistry,
and physical science.

Coastal Conservation Association–Maine
www.ccamaine.org
500 U.S. Route One, Ste. 23
Yarmouth, ME 04096
ph: 207-846-1015
fax: 207-846-1168
tmiller1@ccamaine.org
Focuses on education, legislation, and restoration of marine
fisheries and habitat. CCA is an organization of state chapters composed of avid recreational fishermen who have
banded together to address conservation issues nationally
and within their respective states.

Cobscook Bay Resource Center
www.cobscook.org
4 Favor St.
Eastport, ME 04631
cobscook@prexar.com
Encourages and strengthens community-based approaches
to resource management and sustainable economic development in the Cobscook Bay region, the Bay of Fundy, and
the Gulf of Maine.

Conservation Law Foundation

www.clf.org
14 Maine St., Ste. 200
Brunswick, ME 04011-2026
ph: 207-729-7733
fax: 207-729-7373
mchicoria@clf.org
Works toward solving the most significant environmental problems that threaten the people, natural resources, and communities of New England with a strong focus on marine conservation issues. CLF's advocates use law, economics, and science to create innovative strategies to conserve natural resources, protect public health, and promote vital communities in the region.

East Penobscot Bay Environmental Alliance

jmcclehypernet.com
PO Box 482
Deer Isle, ME 04627
Works to conserve the coast and promote the environmentally appropriate use of East Penobscot Bay. Formed in 2001, EPBEA also defends the bay against the placement of salmon pens near Little Deer Isle. Members are property owners, fishermen, business people, environmentalists, artists, and recreational users of the bay.

Friends of Blue Hill Bay

www.fobhb.org
PO Box 1633
Blue Hill, ME 04614
ph: 207-359-0963
fobhbehypernet.com
Works to preserve the natural ecology, traditional marine fisheries, and the unique aesthetic quality of Blue Hill Bay, particularly against threats posed by salmon farming. This nonprofit organization also supports a baywide management plan that includes all those who have a vested interest in a healthy bay.

Friends of Casco Bay

www.cascobay.org
2 Fort Rd.
South Portland, ME 04106
ph: 207-799-8574
fax: 207-799-7224
keeper@cascobay.org
Works to protect the environmental health of Casco Bay through advocacy, education, collaborative partnerships, and water-quality monitoring. FOCB is also the headquarters of the Casco Baykeeper.

Friends of the Kennebec Salmon

www.kennebecriver.org
PO Box 2473
Augusta, ME 04338
ph: 207-626-8178
fks@gwi.net
Promotes restoration of wild Atlantic salmon to their native home in Maine's Kennebec River.

Friends of Merrymeeting Bay

http://link75.org/mmb/FOMB/fomb.html
PO Box 233
Richmond, ME 04357
ph: 207-721-0941
fax: 207-666-8481
fomb@gwi.net
Seeks to preserve, protect, and improve the unique ecosystems of Merrymeeting Bay on the central coast of Maine through education, research, membership activities, and the promotion and stewardship of conservation easements. Friends is a nonprofit organization founded in 1975.

Friends of Scarborough Marsh

www.scarboroughmaine.com/marsh
PO Box 7049
Scarborough, ME 04070
info_fosm@hotmail.com
Dedicated to the conservation, protection, restoration, and enhancement of Maine's largest saltwater marsh and its watershed. Friends is a coalition of private citizens and organizations working to protect this more than 3,000-acre marsh that acts as a key link to the sea.

Gulf of Maine Research Institute

www.gma.org
PO Box 7549
Portland, ME 04112
ph: 207-772-2321
fax: 207-772-6855
www@gma.org
Works to develop a new genre of research and education institution that combines extensive strategic capacity with a high volume public interface. Formerly the Gulf of Maine Aquarium, GMA teaches about aquatic environments while facilitating marine research.

Island Institute

www.islandinstitute.org
PO Box 648
Rockland, ME 04841
ph: 207-594-9209
fax: 207-594-9314
inquiry@islandinstitute.org

Works for the balanced future of the islands and waters of the Gulf of Maine, from both an environmental and social point of view.

Lobster Conservancy
www.lobsters.org
PO Box 235
Friendship, ME 04547
ph: 207-832-8224
fax: 207-832-8228
dcowan@lobsters.org
Seeks to safeguard the lobster fishery from the crashes that have afflicted many other Gulf of Maine fisheries. The organization works to improve biological knowledge of the American lobster and put that knowledge to work in fishery and environmental management.

Maine Coastal Program
www.state.me.us/spo/mcp
38 State House Station
Augusta, ME 04333
ph: 207-287-3261
fax: 207-287-6489
lorraine.lessard@maine.gov
Works to achieve balance between human uses and the protection of the resources that make Maine's coastal areas so appealing.

Maine Coast Heritage Trust
www.MCHT.org
Bowdoin Mill
1 Main St., Ste. 201
Topsham, ME 04086
ph: 207-729-7366
fax: 207-729-6863
info@mcht.org
Conserves coastal and other lands that define Maine's distinct landscape, protect its environment, sustain its outdoor traditions, and promote the well-being of its people. Since 1970, MCHT has helped permanently protect more than 117,000 acres of wildlife habitat, bold headlands, farmland, and forestland and 241 entire islands. MCHT provides conservation advisory services free of charge to landowners, federal, state and community officials, and local land trusts throughout Maine.

Maine Department of Marine Resources
www.state.me.us/dmr
21 State House Station
Augusta, ME 04333
ph: 207-624-6550
fax: 207-624-6024

Provides leadership in marine policy, management of marine resources, development of sustainable marine resource-based business, and protection of the marine environment.

Maine Lobstermen's Association
www.mainelobstermen.org
1 High St., Ste. 5
Kennebunk, ME 04043
ph: 207-985-4544
fax: 207-985-8099
patrice@mainelobstermen.org
An association of independent lobster fishermen of the Maine coast for the members' mutual benefit and assistance, to maintain and conduct the association, and to promote a program to bring about a more friendly spirit of cooperation among all lobster fishermen.

Maine Maritime Museum
www.bathmaine.com
243 Washington St.
Bath, ME 04530
ph: 207-443-1316
fax: 207-443-1665
maritime@bathmaine.com
Collects, preserves, and interprets material relating to the maritime history of Maine and promotes an understanding and appreciation thereof.

Marine Environmental Research Institute (MERI)
www.meriresearch.org
55 Main St.
PO Box 1652
Blue Hill, ME 04614
ph: 207-374-2135
fax: 207-374-2931
info@meriresearch.org
Promotes scientific research and education on the impacts of pollution on marine life and human health. Through multidisciplinary research, research dissemination, education, and public outreach, MERI strives to protect the health and biodiversity of the marine environment for future generations.

Natural Resources Council of Maine
www.maineenvironment.org
3 Wade St.
Augusta, ME 04330
ph: 207-622-3101
fax: 207-622-4343
nrcm@nrcm.org
Promotes pure water, clean air, vibrant forests, and safety for animals throughout the state.

The Nature Conservancy–Maine
http://nature.org/wherewework/northamerica/states/maine
14 Maine St., Ste. 401
Brunswick, ME 04011
ph: 207-729-5181
fax: 207-729-4118
naturemaine@tnc.org
Seeks to preserve the plants, animals, and natural communities that represent the diversity of life on earth through
land purchases, establishing conservation easements, and
other means. The Nature Conservancy is a national organization headquartered in Arlington, Virginia. In Maine, TNC
has helped preserve thousands of acres of spawning rivers
and coastal wetlands.

Northwest Atlantic Marine Alliance
www.namanet.org
200 Main St., Ste. A
Saco, ME 04072
ph: 207-284-5374
fax: 207-284-1355
craig@namanet.org
Seeks to restore and enhance an enduring Northwest
Atlantic marine system that supports a healthy diversity and
abundance of marine life and human uses, through community-based self-organizing and self-governing institutions.

The Ocean Conservancy–Maine Office
www.oceanconservancy.org
19 Commercial St.
Portland, ME 04101
ph: 207-879-5444
fax: 207-879-5445
john.phillips63@verizon.net
Uses science-based advocacy, research, and public education to inform, inspire, and empower people to speak and
act for the oceans. The Ocean Conservancy is a national
organization dedicated to protecting ocean ecosystems. The
organization has a national office in Washington, D.C., and
ten regional offices, including one in Maine. Its major programs include coral reef protection, debris monitoring,
stormwater runoff and pollution, and coastal cleanup.

Penobscot Bay Watch/Penobscot Bay Alliance
www.penbay.org
PO Box 1871
418 Main St.
Rockland, ME 04841
ph: 207-594-5717
penbay@justice.com
Seeks to protect and monitor the Penobscot Bay area of
Maine.

Sierra Club–Maine Chapter
www.maine.sierraclub.org
One Pleasant St.
Portland, ME 04101-3936
ph: 207-761-5616
fax: 207-773-6690
maine.chapter@sierraclub.org
Works to safeguard Maine's clean water and coastline, and
supports pro-environment candidates for public office. For
over a century the Sierra Club has been devoted to the conservation of forests, mountains, rivers, coasts, and other
natural areas.

Surfrider Foundation–Maine
http://nnesurfriderchapter.org/
PO Box 551
Portland, ME 04112
ph: 207-653-8951
Uses conservation, activism, research, and education to
encourage people to protect and enjoy the world's waves
and beaches. The Surfrider Foundation is a nonprofit environmental organization whose membership consists mainly
of surfers. Its scope has expanded to include protecting
wetlands, bird life, and beaches as well as surf spots. It has
sixty-six chapters worldwide. In addition, Maine members
have no fear of cold water.

MARYLAND
Alliance for the Chesapeake Bay
www.acb-online.org
6600 York Rd., Ste. 100
Baltimore, MD 21212
ph: 410-377-6270
Builds and fosters partnerships to protect and to restore the
bay and its rivers. The alliance works to bring dialogue
between groups that don't see eye-to-eye, forming strategies for joint solutions, and building the capacity of communities for local-level action. It develops methods and
tools for restoration activities and trains citizens to use
them. It mobilizes decision makers, stakeholders, and other
citizens to learn about bay issues and participate in resolv-
ing them.

American Fisheries Society
www.fisheries.org
5410 Grosvenor Ln., Ste. 110
Bethesda, MD 20814
ph: 301-897-8616
fax: 301-897-8096
main@fisheries.org
Works to improve the conservation and sustainability of
fishery resources and aquatic ecosystems by advancing fisheries and aquatic science and promoting the development
of fisheries professionals.

American Zoo and Aquarium Association
www.aza.org
8403 Colesville Rd., Ste. 710
Silver Spring, MD 20910-3314
ph: 301-562-0777
fax: 301-562-0888
generalinquiry@aza.org
Promotes excellence in animal care and welfare, marine
and terrestrial conservation, education, and research that
collectively inspire respect for animals and nature. The association provides its members and their visitors the best possible services through establishing and maintaining excellent professional standards in all AZA institutions, including
ocean aquariums, through its accreditation program and
establishing and promoting high standards of animal care
and welfare.

Assateague Coastal Trust/Coastkeeper
www.actforbays.org
PO Box 731
Berlin, MD 21811
ph: 410-629-1538
fax: 410-629-1059
mail@actforbays.org
Works to preserve Assateague Island and the living
resources of the coastal ecosystem by sponsoring outreach
programs to promote awareness among Delmarva's citizens
and visitors about the natural resources and their long-term
sustainability and participating in advocacy efforts to influence public policies that affect the functions of these
ecosystems.

Audubon Maryland–District of Columbia
www.audubonmddc.org
23000 Wells Point Ln.
Bozman, MD 21612
ph: 410-745-9283
fax: 410-745-9230
Works to conserve and restore natural ecosystems, focusing
on birds, other wildlife, and their habitats for the benefit of
humanity and the earth's biological diversity. Audubon is a
national network of community-based nature centers and
chapters, scientific and educational programs, and advocacy
on behalf of areas sustaining important bird populations.

Calvert Marine Museum
www.Calvertmarinemuseum.com
PO Box 97
Solomons, MD 20688
ph: 410-326-2042
alvscd@co.cal.md.us
Features an educational, regionally oriented museum dedicated to the collection, preservation, research, and interpre-
tation of the culture and natural history of southern
Maryland, its waters and maritime heritage.

Center for Watershed Protection
www.cwp.org
8390 Main St., 2nd Fl.
Ellicott City, MD 21043-4605
ph: 410-461-8323
fax: 410-461-8324
center@cwp.org
Provides local governments, activists, and watershed organizations around the country with the technical tools for protecting some of the nation's most precious natural
resources: streams, lakes, and rivers. Founded in 1992, the
center has developed and disseminated a multidisciplinary
strategy for watershed protection that encompasses watershed planning; watershed restoration; stormwater management; watershed research; better site design, education,
and outreach; and watershed training.

Chesapeake Bay Ecological Foundation
www.chesbay.org
PO Box 1538
Easton, MD 21601
staff@chesbay.org
Strives to educate the public and work with various state
and federal agencies to provide current data on any relevant matters involving the quality of water and the living
resources within and surrounding the Chesapeake Bay.

Chesapeake Bay Foundation
www.cbf.org/site/PageServer?pagename=cbf_homepage
The Philip Merrill Environmental Center
6 Herndon Ave.
Annapolis, MD 21403
ph: 410-268-8816
chesapeake@cbf.org
Works to save the Chesapeake Bay watershed by reducing
pollution, improving fisheries, and protecting and restoring
natural resources such as wetlands, forests, and underwater
grasses. CBF also operates fifteen environmental education
programs. CBF's headquarters are in Annapolis, Maryland,
and it has state offices in Maryland, Virginia, and
Pennsylvania.

Chesapeake Bay Maritime Museum
www.cbmm.org
Navy Point
PO Box 636
St. Michaels, MD 21663
ph: 410-745-2916
fax: 410-745-6088
comments@cbmm.org
Promotes the furthering of interest, understanding, and
appreciation of the culture and maritime heritage of the
Chesapeake Bay and its tributaries through collection, education, documentation, exhibition, research, and publication.

Chesapeake Bay Trust
www.chesapeakebaytrust.org
60 West St., Ste. 405
Annapolis, MD 21401
ph: 410-974-2941
fax: 410-269-0387
postmaster@cbtrust.org
Promotes public awareness and participation in protecting
and restoring Chesapeake Bay and its Maryland tributaries.

Chester River Association
www.chesterriverassociation.org
100 North Cross St., Ste. One
Chestertown, MD 21620
ph: 410-810-7556
fax: 410-810-7555
info@chesterriverassociation.org
Advocates for the health of the Chester River and the living
resources it supports. CRA strives to promote stewardship
of the Chester River—its forests, marshes, fields, creeks,
and streams—as well as an understanding of the river's
place in the economic and cultural life of its communities.
Founded in 1986, CRA is one of the oldest nonprofit watershed associations on the bay, linking all those who live,
work, and play on the banks of the Chester. CRA's corps of
volunteer Chester Testers have created the first long-range
database of the river's water quality, monitoring at
sixteen sites.

Clean Islands International
www.islands.org
8219 Elvanton Dr.
Pasadena, MD 21122-3903
ph: 410-647-2500
fax: 410-647-4554
cii@islands.org
Provides educational and technical assistance to island
communities for the preservation and beautification of
indigenous natural environments and the sustainable development of proper solid waste handling practices. Clean
Islands International is a nonprofit nongovernmental organization with international members.

Environmental Concern Inc.
www.wetland.org
PO Box P
201 Boundry Ln.
St. Michaels, MD 21663
ph: 410-745-9620
fax: 410-745-3517
admin@wetland.org
Promotes wetland education, restoration, and research. EC
is a public nonprofit corporation founded in 1972 by Dr.
Edgar Garbisch who pioneered the technology needed to
construct marshes.

Herring Run Watershed Association
www.herringrun.org
PO Box 24567
Baltimore, MD 21214
ph: 410-254-1577
watershed@herringrun.org
Works to improve the environmental quality of the Herring
Run watershed for the mutual benefit of its community and
the Chesapeake Bay by mobilizing volunteers for advocacy,
restoration, and education.

Khaled bin Sultan Living Oceans Foundation
www.livingoceansfoundation.org
8181 Professional Place, Ste. 215
Landover, MD 20785
ph: 301-577-1288
fax: 301-577-1289
prenaud@livingoceansfoundations.org
Dedicated to conservation and the restoration of living
oceans. Works for their preservation through research, education, and a commitment to science without borders.

**Living on the Edge NOAA's National Ocean Service
Office of Coastal Resource Management**
www.living-edge.org
1305 East-West Hwy., SSMC4
Silver Spring, MD 20910
coastweek@noaa.org
Educates the public about the fragile balance of development and environment along the nation's coastlines. This
group represents a partnership of local and national
organizations.

The Marine Technology Society
www.mtsociety.org
5565 Sterrett Place, Ste. 108
Columbia, MD 21044
ph: 410-884-5330
fax: 410-884-9060
mtsmbrship@erols.com
Disseminates marine science and technical knowledge to
promote and support education for marine scientists, engineers, and technicians. The society advances the development of tools and procedures required to explore, study,
and further the responsible and sustainable use of the
oceans. It provides services that create a broader understanding of the relevance of the marine sciences to other
technologies, arts, and human affairs.

Maryland PIRG–Annapolis
190 Duke of Gloucester St.
Annapolis, MD 21401
ph: 410-267-1160
fax: 208-439-0805
info@marypirg.org

Educates citizens to participate in the political process, including work on fisheries reform and other marine-related issues.

Maryland PIRG–Baltimore

www.marypirg.org
3121 St. Paul St., #26
Baltimore, MD 21218
ph: 410-467-0439
fax: 410-366-2051
Educates citizens to participate in the political process, including work on fisheries reform and other marine-related issues.

Maryland Watermen's Association

www.marylandwatermen.com
1805A Virginia St.
Annapolis, MD 21401
ph: 410-268-7722
fax: 410-268-6635
info@marylandwatermen.com
Promotes the interests of all who derive beauty and benefit from Maryland's Chesapeake Bay Waters. The association's core constituency is made up of commercial fishermen concerned about the pollution of the bay.

National Aquarium in Baltimore

www.aqua.org
Pier 3–501 East Pratt St.
Baltimore, MD 21202
ph: 410-576-3800
conserve@aqua.org
Seeks to stimulate interest in, develop knowledge about, and inspire stewardship of aquatic environments. A member of the Baltimore community, Maryland's leading tourist attraction, and an international icon, the aquarium provides cultural, recreational, and educational experiences that meet the needs of diverse communities.

National Marine Aquaculture Task Force

6930 Carroll Avenue, Ste. 400
Takoma Park, MD 20912
ph: 301-559-0213
chris.mann@earthlink.net
Uses a precautionary approach toward the development and use of marine resources and formulates recommendations to guide decision making on marine aquaculture, particularly in federal waters. NMATF will present a report of its findings to Congress and the public in early 2006.

National Marine Sanctuary Foundation

www.nmsfocean.org
8601 Georgia Ave., Ste. 201
Silver Spring, MD 20910
301-608-3040
fax: 301-608-3044
lori@mnsfocean.org

Creates meaningful opportunities for public interaction with the nation's marine sanctuaries. Through public and private sector partnerships, the foundation runs conservation-based education and outreach programs designed to preserve, protect, and promote these underwater treasures.

National Marine Sanctuary Program–NOAA

www.sanctuaries.nos.noaa.gov
1305 East-West Hwy., 11th Fl.
Silver Spring, MD 20910
ph: 301-713-3125
fax: 301-713-0404
sanctuaries@noaa.gov
National office of the fourteen U.S. Marine Sanctuaries (see Marine Parks section).

Oceanography Society

www.tos.org
PO Box 1931
Rockville, MD 20859-1931
ph: 301-251-7708
fax: 301-251-7709
info@tos.org
Disseminates knowledge of oceanography and its application through research and education, promotes communication among oceanographers, and provides a constituency for consensus building across all the disciplines of the field.

Oyster Recovery Partnership

www.oysterrecovery.org
PO Box 6775
Annapolis, MD 21401
ph: 410-990-4970
orp@oysterrecovery.org
Restores the ecological and economic benefits of oysters to Chesapeake Bay by creating habitat through rehabilitation of historically productive oyster bars and seeding those bars with hatchery disease-free oysters produced at the University of Maryland's Center for Environmental Science Horn Point Lab.

Sierra Club–Maryland

www.maryland.sierraclub.org
7338 Baltimore Ave., Ste. 101A
College Park, MD 20740
ph: 301-277-7111
fax: 301-277-6699
laurel.imlay@sierraclub.org
Uses lawful means to practice and promote the responsible use of the earth's ecosystems and resources and to educate and enlist humanity to protect and restore the quality of the natural and human environment. The Maryland Sierra Club is working to control factory farms and other threats to the water quality of the bay.

South River Federation/South Riverkeeper
www.southriverfederation.net
6 Herndon Ave.
Annapolis, MD 21403
ph: 443-482-2155
cindy@southriverfederation.net
Seeks to preserve, protect, and restore the South River and
its ecosystem.

Surfrider Foundation–Maryland, Ocean City Chapter
www.surfrider.org/oceancitymd
PO Box 3342
Ocean City, MD 21843
ph: 410-956-2199
oceancitymd@surfrider.org
Uses conservation, activism, research, and education to
encourage people to protect and enjoy the world's waves
and beaches. The Surfrider Foundation is a nonprofit
environmental organization whose membership consists
mainly of surfers. Its scope has expanded to include protecting wetlands, bird life, and beaches as well as surf
spots. It has sixty-six chapters worldwide.

U.S. Naval Academy
www.usna.edu
121 Blake Rd.
Annapolis, MD 21402-5000
ph: 410-293-1000
fax: 410-293-1000
Provides young men and women the up-to-date academic
and professional training needed to be effective naval and
marine officers after their graduation. The Naval Academy
was founded in 1845.

MASSACHUSETTS
Association for the Preservation of Cape Cod
www.apcc.org
The Barnstable House
3010 Main St.
PO Box 398
Barnstable, MA 02630
ph: 508-362-4226
fax: 508-362-4227
info@apcc.org
Works to preserve the quality of life on Cape Cod.

Audubon Society–Massachusetts
www.massaudubon.org
208 South Great Rd.
Lincoln, MA 01773
ph: 781-259-9500
webmaster@massaudubon.org
Works to protect the nature of Massachusetts for people
and wildlife, including the coast and its shorebirds. The
Massachusetts Audubon Society is the largest conservation
organization in New England. Audubon's mission is to
conserve and restore natural ecosystems, focusing on birds,
other wildlife, and their habitats for the benefit of humanity
and the earth's biological diversity.

Boston Harbor Association
www.tbha.org
374 Congress St., Ste. 609
Boston, MA 02210
ph: 617-482-1722
fax: 617-482-9750
mail@tbha.org
Works to promote a clean, alive, and accessible Boston
Harbor.

Buzzards Baykeeper/Coalition for Buzzards Bay
www.savebuzzardsbay.org
620 Belleville Ave.
New Bedford, MA 02745
ph: 508-999-6363
fax: 508-984-7913
cbb@savebzzardsbay.org
Work throughout the entire Buzzards Bay watershed to
protect the region's coastal, river, and drinking water quality and the upland forests, wetlands, and streams that support a healthy watershed and bay ecosystem.

Cape Cod Center for Sustainability
244 Willow St.
Yarmouth Port, MA 02675
ph: 508-362-6313
Studies and promotes environmentally compatible economic
development, encourages ecological tourism, and develops
demonstration models for the Cape community.

**Cape Cod Commercial Hook Fisherman's
Association**
www.ccchfa.org
210 Orleans Rd., Box 2
North Chatham, MA 02650
ph: 508-945-2432
fax: 508-945-0981
contact@ccchfa.org
Works toward the future to protect a resource, a tradition,
a way of life. This alliance of fishermen, local residents, and
environmentalists seeks ways to maintain an affordable
and sustainable livelihood for area fishermen and the
resource they depend on.

Cape Cod Commission
www.capecodcommission.org
PO Box 226
Barnstable, MA 02630
ph: 508-362-3828
fax: 508-362-3136
frontdesk@capecodcommission.org

Seeks to protect the shore and the Cape. This is the regional planning and land-use regulatory agency.

Cape Cod Stranding Network
www.CapeCodStranding.net
PO Box 287
Buzzards Bay, MA 02532
ph: 508-743-9805
fax: 508-759-5477
ccsninfo@capecodstranding.net
Responds to stranded marine mammals along a 700-mile coastline.

Center for Coastal Studies
www.coastalstudies.org
PO Box 1036
Provincetown, MA 02657
ph: 508-487-3622
fax: 508-487-4495
ccs@coastalstudies.org
Conducts scientific research, promotes stewardship of resources, and provides educational activities with emphasis on the coastal and marine habitats and resources of the Gulf of Maine.

Center for Oceanic Research and Education
www.coreresearch.org
245 Western Ave.
Essex, MA 01929
ph: 978-768-4560
fax: 978-768-4560
core@coreresearch.org
Features whale-watching education programs, school presentations, and public outreach to educate people about endangered whales and their fragile marine environment. CORE is a nonprofit organization dedicated to the study and conservation of cetaceans.

Charles River Conservancy
www.charlesriverconservancy.org
c/o EF Education
EF Center Boston
One Education St.
Cambridge, MA 02141
ph: 617-619-2850
fax: 617-619-2856
crc@charlesriverconservancy.org
CRC is a nonprofit advocacy group founded in 1999 dedicated to the renovation, maintenance, and enrichment of the Charles River Basin and its surroundings, particularly its parks, parkways, and bridges.

Charles River Watershed Association
www.crwa.org
48 Woerd Ave., Ste. 103
Waltham, MA 02453
ph: 781-788-0007
fax: 781-788-0057
rzimmerman@crwa.org
Protects the Charles River and its watershed.

Citizens for the Protection of Waquoit Bay
www.waquoitbayreserve.org/friends.htm
PO Box 3407
Waquoit, MA 02536
ph: 508-540-1948
Strives to protect Waquoit Bay National Estuarine Research Reserve.

Conservation Law Foundation
www.clf.org
62 Summer St.
Boston, MA 02110-1016
ph: 617-350-0990
fax: 617-350-4030
achatwin@clf.org
Works to solve environmental problems that threaten the people, natural resources, and communities of New England, with a strong focus on marine protection.

Earthwatch Institute
www.earthwatch.org
PO Box 75
3 Clock Tower Place, Ste. 100
Maynard, MA 01754
ph: 978-461-0081
fax: 978-461-2332
info@earthwatch.org
Creates partnerships to promote conservation, research, and education, including on behalf of global seas.

Environmental League of Massachusetts
www.environmentalleague.org
14 Beacon St., Ste. 714
Boston, MA 02108
ph: 617-742-2553
fax: 617-742-9656
elm@environmentalleague.org
Advocates responsible environmental policy on a range of issues, including coastal protection and marine water quality. The league was founded in 1898.

Friends of Pleasant Bay
www.fopb.org
PO Box 845
South Orleans, MA 02662
houston@harwich.edu
Preserves Cape Cod's recreational and ecological resources.

Gloucester Fishermen's Wives Association
www.gfwa.org
11-15 Parker St.
Gloucester, MA 01930
ph: 978-282-1401
fax: 978-283-7304
info@gfwa.org
Promotes the New England fishing industry and quality of life for active and retired fisherman. Also seeks to preserve the habitat of the ocean.

Gulf of Maine Council on the Marine Environment–Massachusetts Coastal Zone Management
www.gulfofmaine.org
251 Causeway St., Ste. 900
Boston, MA 02114-2151
ph: 617-626-1202
fax: 617-626-1240
susan.snow-cotter@state.ma.us
Promotes cooperation to maintain and enhance the environmental quality of the marine waters of the Gulf of Maine.

Hands Across the River Coalition
http://wetlandsandwildlife.com:85/wickedneat/harc/mission.htm
220 Union, Rm. 202
New Bedford, MA 02740
harcnb@aol.com
Fights historic pollution of the river and works to save wetlands to protect the future.

International Fund for Animal Welfare
www.ifaw.org
411 Main St.
PO Box 193
Yarmouth Port, MA 02675
ph: 508-744-2000
fax: 508-744-2009
info@ifaw.org
Engages communities, government leaders, and like-minded organizations around the world in an effort to achieve lasting solutions to pressing animal welfare and conservation challenges and solutions that benefit both animals and people. This office has a particular focus on whale conservation.

Ipswich River Watershed Association
www.ipswichriver.org
PO Box 576
Ipswich, MA 01938
ph: 978-887-2313
fax: 978-887-2208
irwainfo@ipwichriver.org

Seeks to protect the Ipswich River and its contributing watershed. The association is a group of concerned individuals, families, businesses, officials, educators, scientists, and managers.

Island Alliance
www.bostonislands.org
408 Atlantic Ave., Ste. 228
Boston, MA 02110
ph: 617-223-8530
fax: 617-223-8671
info@islandalliance.org
Supports the Boston Harbor Islands national park area by attracting investment and support for the park and its mission.

IWC and the Whale Adoption Project
www.iwc.org
70 E Falmouth Highway
East Falmouth, MA 02536
ph: 508-548-8328
fax: 508-457-1988
iwchq@iwc.org
Saves endangered species and protects wildlife habitat through public education and other efforts.

Jason Foundation for Education
www.jason.org
11 Second Ave.
Needham Heights, MA 02494
ph: 781-444-8858
fax: 781-444-8313
info@jason.org
Provides experience-based curriculum and professional development for science educators with an emphasis on oceans.

Manomet Center for Conservation Sciences
www.manomet.org
81 Stage Rd.
PO Box 1770
Manomet, MA 02345
ph: 508-224-6521
fax: 508-224-9220
Conducts original research on natural systems and wildlife to bring people together and guide them in developing practical strategies that improve conditions for marine and terrestrial wildlife, habitats, and people.

Marine Education Center of Cape Ann
www.cape-ann.com/7seas/whalewatch/links.html
PO Box 3015
Gloucester, MA 01930
ph: 978-546-9538
fax: 978-546-9538
flipper@cove.com

Promotes awareness of coastal ocean life through education, conservation, and research. Many marine-related educational activities can be found at MECCA, including an adopt-a-whale program.

Mashpee Environmental Coalition

www.mashpeemec.us
PO Box 274
Mashpee, MA 02649
alomashpee@aol.com
Provides a grassroots community problem-solving infrastructure disseminating early identification and accurate threat analysis information.

Massachusetts Bay Marine Studies Consortium

www.brandeis.edu/marinestudies/index.html
900 Washington St.
Wellesley, MA 02482-5275
ph: 781-444-3643
rjstern@bsn1.net
Strives to increase understanding of coastal and marine ecosystems and to promote policies that protect aquatic resources, especially through reduction of nonpoint source pollution. The consortium is an association of Massachusetts higher education and research institutions.

Massachusetts Bays National Estuaries Program

www.mass.gov/envir/massbays
251 Causeway St., Ste. 800
Boston, MA 02411
massbays@state.ma.us
Strives to protect and enhance coastal health and heritage of Massachusetts and Cape Cod bays. This group is a partnership of citizens, communities, and government.

Massachusetts Environmental Trust

www.agmconnect.org/massenvironmentaltrust/
met-home.htm
100 Cambridge St., Ste. 1060
Boston, MA 02114
ph: 617-626-1059
fax: 617-626-1059
Protects and preserves the Commonwealth's water resources and its ecosystems through grants to conservation efforts. Grants are funded primarily by the purchase of special environmental license plates. Supported efforts include clean water, watershed protection, and protection of endangered northern right whales.

Massachusetts Land Trust Coalition

www.massland.org
18 Wolbach Rd.
Sudbury, MA 01776
ph: 978-443-5588
fax: 978-443-2333
info@massland.org

Increases the effectiveness of land trusts and conservation organizations in Massachusetts by working with the legislature and governmental agencies on issues of direct interest to the conservation movement.

Massachusetts PIRG

www.masspirg.org
44 Winter St., 4th Fl.
Boston, MA 02108
ph: 617-292-4800
fax: 617-292-8507
info@masspirg.org
Educates citizens to participate in the political process. Works on fisheries and other marine issues.

Massachusetts Society of Conservation Biology

www.nescb.org
263 South Ave.
Weston, MA 02493
www@nescb.org
Promotes environmental planning, ensures that communities are adequately informed, and focuses on the conservation of species, habitats, and vital ecosystem services by targeting specific issues from local to global scales for concerted intervention. The society includes New England conservation researchers, educators, students, and practitioners.

Massachusetts Water Watch

www.masswwp.org
Blaisdell House
University of Massachusetts
Amherst, MA 01003
ph: 413-545-5531
fax: 413-545-2304
mwwp@tei.umass.edu
Provides training and other technical assistance to citizen organizations that conduct water quality monitoring programs on the lakes, rivers, and estuaries of Massachusetts.

Massachusetts Wildlife Federation

www.masswildfed.org
PO Box 188
Concord, MA 01742
info@masswildfed.org
Promotes conservation of wildlife and wildlife habitat through comprehensive natural resource management and a variety of compatible activities. The federation promotes and offers advice on educational and legislative programs, addressing issues that affect wildlife habitat and environmental quality primarily, but not exclusively, in Massachusetts.

Merrimack River Watershed Council

www.merrimack.org
600 Suffolk St., 4th Fl.
Lowell, MA 01854
ph: 978-275-0120
fax: 978-275-0125
ecoughlin@merrimack.org
Seeks to protect the natural resources of the watershed.
MRWC is a not-for-profit membership organization with a
growing constituency of individuals, businesses, municipali-
ties, and community groups.

National Environmental Law Center

www.nelconline.org
44 Winter St., 4th Fl.
Boston, MA 02108
ph: 617-422-0880
fax: 617-422-0881
nelc@nelconline.org
Works to enforce anti-pollution laws and promotes long-
term solutions to the nation's most pressing environmental
problems. NELC scientists, lawyers, and policy experts have
a proven track record of bringing corporate polluters to justice and translating innovative ideas into practical reforms.
NELC has successfully sued the Maine salmon farming
industry, arguing they negatively impact wild salmon and
coastal habitat.

The Nature Conservancy–Massachusetts Chapter

www. nature.org/wherewework/northamerica/states/
massachusetts/
205 Portland St., Ste. 400
Boston, MA 02114-1708
ph: 617-227-7017
fax: 617-227-7688
Seeks to preserve the plants, animals, and natural communities that represent the diversity of life on earth through
land purchases, establishing conservation easements, and
other means. The Nature Conservancy is a national organization headquartered in Arlington, Virginia. In
Massachusetts TNC is working to protect natural areas on
Martha's Vineyard, Nantucket, and surrounding islands.

New Bedford Whaling Museum

www.whalingmuseum.org
18 Johnny Cake Hill
New Bedford, MA 02740-6398
ph: 508-997-0046
fax: 508-997-0018
abrengle@whalingmuseum.org
Seeks to educate and interest the public in the historic
interaction of humans with whales worldwide, the history
of Old Dartmouth and adjacent communities, and regional
maritime activities.

New England Aquarium

www.neaq.org
Central Wharf
Boston, MA 02110-3399
ph: 617-973-5200
Provides an underwater experience for visitors and a cultur-
al institution that reconnects Boston to its waterfront. NEA
is one of the best known and most visited aquariums in the
United States.

New England Campaign for a Healthy Ocean

www.clf.org
c/o Conservation Law Foundation
62 Summer St.
Boston, MA 02110-1016
ph: 617-350-0990
fax: 617-350-4030
Works to correct causes that have brought several whale
species to near-extinction and have closed many historically
important fisheries, resulting in devastation to coastal communities long dependent upon the Gulf's legendary marine
wildlife and fisheries.

Ocean Alliance

www.oceanalliance.org/wci
191 Weston Rd.
Lincoln, MA 01773
ph: 800-969-4253
fax: 781-259-0288
question@oceanalliance.org
Promotes conservation of whales and their ocean environment through research and education.

Parker River Clean Water Association

www.parker-river.org
PO Box 798
Byfield, MA 01922
ph: 978-462-2551
info@parker-river.org
Promotes restoration and protection of the waters and
environment of the Parker River and Plum Island Sound
watersheds.

Salem Sound Coastwatch

www.salemsound.org
201 Washington St., Ste. 9
Salem, MA 01970
ph: 978-741-7900
info@salemsound.org
Works with government agencies, businesses, other nonprofit organizations, and citizens from the communities of
Manchester, Beverly, Danvers, Peabody, Salem, and
Marblehead to take cooperative action to protect and
enhance the environmental quality of the Salem Sound
watershed.

Save the Harbor/Save the Bay
www.savetheharbor.org
286 Congress St., 7th Fl.
Boston, MA 02210
ph: 617-451-2860
fax: 617-451-0496
Seeks to restore and protect Boston Harbor and
Massachusetts Bay, to reconnect citizens and communities
to the sea, to build the next generation of youth, and to
ensure development of a world-class waterfront for every-
one to enjoy.

Sea Education Association
www.sea.edu
PO Box 6
Woods Hole, MA 02543
ph: 800-552-3633
fax: 508-457-4673
Prepares students for the challenges they will face in an
increasingly complex society. SEA is committed to providing
opportunities for students to gain deeper insights into the
ocean planet by studying the ocean environment through
hands-on experience aboard a traditional sailing ship.

**Sierra Club–Massachusetts, Coastal and Marine
Habitat and Wildlife Committee**
www.sierraclubmass.org
100 Boylston St.
Boston, MA 02116
ph: 617-423-5775
fax: 617-423-5858
webmaster@sierraclubmass.org
Focuses on protecting the state's coastal and marine
resources. The Sierra Club uses lawful means to practice
and promote the responsible use of the earth's ecosystems
and resources and to educate and enlist humanity to protect and restore the quality of the natural and human
environment.

Stellwagen Bank National Marine Sanctuary
http://woodshole.er.usgs.gov/project-pages/stellwagen
175 Edward Foster Rd.
Scituate, MA 02066
ph: 781-545-8026
fax: 781-545-8026
webmaster@usgs.gov
Provides essential habitat for many species of marine mammals, including the endangered North Atlantic right whale,
and is the focus of a large tourism industry centered on
whale watching. The sanctuary region, a rich commercial
and recreational fishing ground, is heavily used by humans
and marine species.

Surfrider Foundation–Massachusetts
www.surfrider.org/massachusetts
Astor Station
PO Box 462
Boston, MA 02123-0462
massachusetts@surfrider.org
Uses conservation, activism, research, and education to
encourage people to protect and enjoy the world's waves
and beaches. The Surfrider Foundation is a nonprofit environmental organization whose membership consists mainly
of surfers. Its scope has expanded to include protecting
wetlands, bird life, and beaches as well as surf spots. It has
sixty-six chapters worldwide.

SWIM (Safer Waters in MA)
www.nahant.org/swim.html
c/o Northeastern U. Marine Science Center/East Pt.
Nahant, MA 01908
ph: 781-581-7370 x311
nahantswim@aol.com
Works to protect the waters of the North Shore from pollution, including both Boston Harbor and Massachusetts Bay.

Three Bays Preservation Society
www.3bays.org
PO Box 215
Osterville, MA 02655
ph: 508-420-0780
fax: 508-420-4489
info@3bays.org
Preserves, maintains, protects, and enhances the aquatic
environment and related ecosystems of the three-bay estuary composed of West Bay, North Bay, and Cotuit Bay and
environs in Barnstable County, Cape Cod, Massachusetts.
The society also acts to forestall and minimize threats to
the health of the Three Bays system.

Trout Unlimited–Massachusetts Council
www.tu.org
65 East St.
Lenox, MA 01240
ph: 413-637-1818
jg_chague@hotmail.com
Strives to conserve, protect, and restore North America's
trout and salmon fisheries and their watersheds. It includes
two chapters in Massachusetts.

Trout Unlimited–Cape Cod Chapter
13 Long Pond Dr.
Harwich, MA 02645
ph: 508-432-8148
jrjlbliss@earthlink.net

Trout Unlimited–Greater Boston Chapter
www.tu.org
436 Webster St.
Needham, MA 02494
ph: 781-455-6905
fax: 781-453-8008
morellbox@aol.com

Urban Ecology Institute
www.urbaneco.org
355 Higgins Hall
140 Commonwealth Ave.
Boston College
Chestnut Hill, MA 02467
ph: 617-552-0592
fax: 617-552-1198
info@urbaneco.org
Offers studies focusing on the trees, rivers, ocean harbors, wildlife, and open spaces found in cities to understand the extent of those resources and the way they are affected by pollution, overdevelopment, and other pressures, and using this information to inform urban planning and policy. Urban ecology is a new branch of environmental studies that seeks to understand the natural systems of urban areas and the threats that face them.

Whale Conservation Institute
www.oceanalliance.org/wci/
191 Weston Rd.
Lincoln, MA 01773
ph: 781-259-0423
fax: 781-259-0288
question@oceanalliance.org
Strives to protect whales and their ocean environment through scientific research and education.

Woods Hole Oceanographic Institution
www.whoi.edu
266 Woods Hole Rd.
Woods Hole, MA 02543
ph: 508-548-1400
fax: 508-457-2034
information@whoi.edu
Promotes the study of marine science and the education of marine scientists. Woods Hole is the largest independent oceanographic institute in the world.

MINNESOTA
Institute for Agriculture and Trade Policy
www.iatp.org
2105 First Ave. S
Minneapolis, MN 55404
ph: 612-870-0543
fax: 612-870-4846
iatp@iatp.org

Promotes resilient family farms, local communities, and ecosystems around the world through research, education, science, technology, and advocacy, including advocating for reduced nutrient and agricultural chemical runoff that cause algae blooms and dead zones in coastal waters.

Mississippi River Basin Alliance
www.mrba.org
708 North First St., Ste. 238
Minneapolis, MN 55401
ph: 612-334-9460
fax: 612-340-1632
mrbaoffice@mrba.org
Works to protect and restore the ecological, economic, cultural, historical, and recreational resources in the basin and to eliminate barriers of race, class, and economic status that divide supporters in the quest to achieve their purposes. These include greatly reducing the pollutants that flow down the river into the Gulf of Mexico, which creates a massive dead zone every spring and summer.

MISSISSIPPI
Audubon Mississippi
www.msaudubon.com
1208 Washington St.
Vicksburg, MS 39183
ph: 601-661-6189
Works to conserve and restore natural ecosystems by focusing on birds, other wildlife, and their habitats for the benefit of humanity and the earth's biological diversity. Audubon is a national network of community-based nature centers and chapters, scientific and educational programs, and advocacy on behalf of areas sustaining important bird populations. Among its works Mississippi Audubon recently produced the first coastal birding trail map in the state's history.

Mississippi–Alabama Sea Grant Consortium
www.masgc.org
PO Box 7000
Ocean Springs, MS 39566-7000
ph: 228-818-8836
fax: 228-818-8841
Promotes activities that foster the conservation and sustainable development of coastal and marine resources in Mississippi and Alabama. The MASGC is a federal-state partnership created in 1972.

Mississippi Wildlife Federation
www.mswildlife.org
855 S. Pear Orchard Rd, Ste. 500
Ridgeland, MS 39157
ph: 601-206-5703
fax: 601-206-5705
cshropshire@mswf.org
Seeks protection of wildlife in Mississippi.

National Marine Educators Association
www.marine-ed.org
PO Box 1470
Ocean Springs, MS 39566
Provides a focus for marine and aquatic studies worldwide.
NMEA brings together those interested in the study and
enjoyment of the world of water—both freshwater and
saltwater.

The Nature Conservancy–Mississippi Chapter
www.nature.org/mississippi
964 North Jefferson St.
Jackson, MS 39202
ph: 601-713-3355
Works to find, protect, and maintain the best examples of
natural communities, ecosystems, and endangered species
in Mississippi. While the national organization has been
active in Mississippi since the 1960s, the chapter office was
founded in 1989. Today, the chapter has twenty employees
in four locations around the state: Jackson, the Mississippi
Gulf Coast, Tupelo, and Camp Shelby. The Nature
Conservancy is a national organization headquartered in
Arlington, Virginia. Through land purchases, establishing
conservation easements and other means, it seeks to preserve the plants, animals, and natural communities that
represent the diversity of life on earth.

Sierra Club–Mississippi Chapter
www.mississippi.sierraclub.org
PO Box 4335
Jackson, MS 39296-4335
ph: 601-352-1026
Uses lawful means to practice and promote the responsible
use of the earth's ecosystems and resources, and to educate and enlist humanity to protect and restore the quality
of the natural and human environment.

Sierra Club–Gulf Coast Group
www.mississippi.sierraclub.org
ph: 228-872-3457
information@sierraclub.org

NEW HAMPSHIRE
Audubon Society–New Hampshire
www.nhaudubon.org
3 Silk Farm Rd.
Concord, NH 03301
ph: 603-224-9909
fax: 603-226-0902
asnh@nhaudubon.org
Promotes the conservation of wildlife and habitat throughout the state. Independent of the National Audubon
Society, ASNH offers programs in wildlife conservation,
land protection, environmental policy, and environmental
education.

Blue Ocean Society for Marine Conservation
www.blueoceansociety.org
370 Portsmouth Ave., #9
Greenland, NH 03840
ph: 603-431-0260
info@blueoceansociety.org
Promotes conservation of the marine environment through
education and research in New England.

Clean Water Action/Clean Water Fund
www.cleanwateraction.org/nh
163 Court St.
Portsmouth, NH 03801
ph: 603-430-9565
fax: 603-430-9708
portcwa@cleanwater.org
Works for clean, safe, and affordable water, prevention of
health-threatening pollution, creation of environmentally
safe jobs and businesses, and empowerment of people to
make democracy work. Clean Water Action is a national citizens' organization.

Great Bay Coast Watch
www.gbcw.unh.edu
Sea Grant, Kingman Farm/UNH
Durham, NH 03824-3512
ph: 603-749-1565
fax: 603-743-3997
ann.reid@unh.edu
Protects New Hampshire's coastal waters through monitoring water quality and educational outreach efforts. This
group is a volunteer organization, run through the UNH
Sea Grant program.

Gulf of Maine Council on the Marine Environment
www.gulfofmaine.org
PO Box 3019
Boscawen, NH 03303-3019
ph: 603-796-2615
fax: 603-796-2600
info@gulfofmaine.org
Works to maintain and enhance environmental quality in
the Gulf of Maine to allow for sustainable resource use by
existing and future generations. The council is a
U.S.–Canadian partnership of government and nongovern-
ment organizations.

New Hampshire PIRG
www.nhpirg.org
80 North Main St.
Concord, NH 03301
ph: 603-229-3222
fax: 603-229-3221
nhpirg@pirg.org
Educates citizens to participate in the political process,
including fisheries reform and other marine issues.

New Hampshire Rivers Council
www.nhrivers.org
54 Portsmouth St.
Concord, NH 03301
ph: 603-228-6472
fax: 603-228-6472
info@nhrivers.org
Promotes the protection and conservation of New
Hampshire's rivers.

Seacoast Anti-Pollution League
www.sapl.org
PO Box 1136
Portsmouth, NH 03802
ph: 603-431-5089
info@sapl.org
Preserves public health and environmental quality in the
seacoast region of New Hampshire.

Seacoast Science Center
www.seacentr.org
570 Ocean Blvd.
Rye, NH 03870
ph: 603-436-8043
fax: 603-433-2235
Offers programs year-round including naturalist-guided and
unguided explorer programs for schools, visitor programs
on the natural and social history of Odiorne Point and the
Gulf of Maine, and a variety of indoor exhibits highlighting
some of the fascinating features of coastal New Hampshire.
The center is situated on 136 acres, where visitors can
explore seven distinct coastal habitats.

Sierra Club–New Hampshire Chapter
www.nhsierraclub.org
40 North Main St., 2nd Fl.
Concord, NH 03301
ph: 603-224-8222
fax: 603-224-4719
feedback@nhsierraclub.org
Uses lawful means to practice and promote the responsible
use of the earth's ecosystems and resources. The New
Hampshire chapter is divided into four geographic
groups—Seacoast, Merrimack Valley, Monadnock, and
Upper Valley—and fights to protect a range of habitats
including the seacoast and marine waters.

Adventure Aquarium–Academy for Aquatic Sciences
www.njaquarium.org
1 Riverside Dr.
Camden, NJ 08103-1060
ph: 856-365-3300
fax: 856-365-3311
Features "bigger, better, wetter" exhibits along with an
ongoing community education and outreach program
through the Academy for Aquatic Sciences. The aquarium
will reopen in spring 2005.

Alliance for a Living Ocean
www.livingocean.org
2007 Long Beach Blvd.
North Beach Haven, NJ 08008
ph: 609-492-0222
fax: 609-492-6216
livingoceanalo@comcast.net
Promotes and maintains clean water and a healthy coastal
environment through education, research, and active
participation.

American Littoral Society–Headquarters
www.littoralsociety.org
Sandy Hook
Highlands, NJ 07732
ph: 732-291-0055
als@netlabs.net
Draws attention to issues that affect the littoral zone—that
area on the beach between low and high tide. The society
is a national nonprofit, public-interest organization composed of over six thousand professional and amateur naturalists, with headquarters in Sandy Hook, New Jersey. It
gets its name from the Latin *litus* meaning beach or
coastal.

American Littoral Society–Northeast Chapter
www.alsnyc.org
PO Box 1306
Tuckerton, NJ 08087
ph: 609-294-3111

Barnegat Bay Watershed and Estuary Foundation
www.bbwef.org
PO Box 364
Island Heights, NJ 08732-0364
ph: 732-505-3671
bbwef@bbwef.org
Promotes a healthy ecosystem throughout the Barnegat
Bay watershed. To accomplish this, the foundation works
toward conservation and protection of soil, forest, woods,
water, shorelines, and fish and wildlife, promoting restoration and conservation of all the natural resources of the
watershed.

Clean Ocean Action

www.cleanoceanaction.org
PO Box 505
Sandy Hook, NJ 07732
ph: 732-872-0111
fax: 732-872-8041
sandyhook@cleanoceanaction.org
Works to improve the degraded water quality of the marine waters off the New Jersey–New York coast. Clean Ocean Action identifies the sources of pollution and attacks each one by using research, public education, and citizen action to convince public officials to enact and enforce measures that will clean up and protect the ocean.

Clean Ocean Action–Wildwood Office

PO Box 1098
Wildwood, NJ 08260
ph: 609-729-9262
fax: 609-729-1091
wildwood@cleanoceanaction.org

Coastal Ocean Coalition

www.coastaloceancoalition.org
PO Box 73
Atlantic Highlands, NJ 07716
ph: 732-291-2163
fax: 732-875-1174
info@coastaloceancoalition.org
Draws attention to the environmental effects of the proposed "Freedom to Fish Act" on efforts to protect and restore coastal and ocean resources. Five regional and national organizations formed the COC in 2003.

Fishers Island Conservancy Inc.

www.seaweb.org/resources/seawebmap/nj.html
PO Box 132
Green Village, NJ 07935
Seeks to preserve and enhance the natural resources of the island, located off New York's Long Island, and its surroundings. Forty endangered species inhabit the surrounding area. Special activities include environmental programs (nontoxic mosquito control) and public speaking events.

Garden Club of Long Beach Island

http://lbinet.com/garden
PO Box 85
Barnagat Light, NJ 08008-0085
ph: 609-494-2851
fax: 609-494-2851
cfoam@att.net
Maintains the gardens at the Barnegat Light Museum.

Hackensack Riverkeeper

www.hackensackriverkeeper.org
231 Main St.
Hackensack, NJ 07601
ph: 201-968-0808
fax: 201-968-0336
info@hackensackriverkeeper.org
Protects and defends the environmental quality of the ecosystem of the estuary, river, and watershed and the quality of life for the people and other creatures that inhabit the Hackensack River watershed. The Hackensack Riverkeeper provides representation for the natural living resources of the Hackensack River, which are manifested in environmental advocacy, education, and conservation programs.

**Hudson River Fishermen's Association–
New Jersey Chapter**

www.hrfanj.org
PO Box 421
Cresskill, NJ 07626
ph: 877-HRFAORG
fax: 201-265-9270
webmaster@hrfanj.org
Encourages responsible use of aquatic resources and protection of habitat. The association is a group of recreational fishermen who make active use of the New York Bight and the surrounding water system and are concerned with the present and future state of these fisheries. The group assists where possible in efforts to abate pollution and promote and manage sportfishing.

Institute of Coastal Education

www.cleanoceanaction.org
PO Box 1098
3419 Pacific Ave.
Wildwood, NJ 08260
ph: 609-729-9262
fax: 609-729-1091
wildwood@cleanoceanaction.org
Features a vibrant interactive center for coastal environmental education and advocacy. With twelve large aquariums holding locally collected species of fish, crustaceans, octopi, and coral, ICE has become a hub of activity for students, teachers, and anyone interested in the coastal habitat of southern New Jersey. The center also offers hands-on activity centers and work stations as well as field trips. This is a joint project the American Littoral Society and Clean Ocean Action.

Jenkinson's Aquarium
www.jenkinsons.com/aquarium
Point Pleasant Beach, NJ 08742
ph: 732-899-1212
jenkinsonsaquarium@comcast.net
Educates the public on all aspects of marine life and conservation. Each exhibit is designed to promote awareness of the animals, their habitats, and conservation. The aquarium is a privately owned facility located on the boardwalk in Point Pleasant Beach.

Jersey Coast Anglers Association
www.jcaa.org
1201 Route 37 East, Ste. 9
Toms River, NJ 08753
ph: 732-506-6565
fax: 732-506-6975
tfote@jcaa.org
Represents the position of marine sport anglers, champions their causes, and protects their rights in matters pertaining to fishing, fisheries, and environmental quality. JCAA, an association of more than seventy-five saltwater fishing clubs, with a combined membership exceeding 30,000, often works in concert with major environmental organizations.

Marine Mammal Stranding Center
www.mmsc.org
3625 Brigantine Blvd.
Brigantine, NJ 08203
ph: 609-266-0538
fax: 609-266-6300
mmsc@verizon.net
Responds to calls for stranded whales, dolphins, seals, and sea turtles that washed ashore on New Jersey beaches. The MMSC, founded in 1978, is a private nonprofit organization.

National Association of Underwater Instructors–Northeast Regional Office
www.naui.org
1333 Curtis Ave.
Point Pleasant, NJ 08742
ph: 904-237-1468
Supports and promotes dive safety through education. Formed in 1960, NAUI Worldwide offers a full range of training programs from skin diver through instructor course director, with dozens of specialty courses including nitrox and technical diving.

New Jersey Audubon Society
www.njaudubon.org
11 Hardscrabble Rd.
PO Box 693
Bernardsville, NJ 07924
ph: 908-766-5787
fax: 908-776-7775
hq@njaudubon.org
Works to conserve and restore natural ecosystems, focusing on birds, other wildlife, and their habitats for the benefit of humanity and the earth's biological diversity. Among Audubon's local efforts is the shorebird/horseshoe crab conservation campaign that targets the overharvesting of horseshoe crabs. The horseshoe crab's population decline also threatens nine species of shorebirds that feed on their eggs.

New Jersey Conservation Foundation
www.njconservation.org
Bamboo Brook
170 Longview Rd.
Far Hills, NJ 07931
ph: 908-234-1225
fax: 908-234-1189
info@njconservation.org
Protects strategic lands, promotes strong land use policies, and forges partnerships to achieve conservation goals. Since 1960, NJCF has protected tens of thousands of acres of open space, from the Highlands to the Pine Barrens to Delaware Bayshore.

New Jersey Council of Diving Clubs
www.scubanj.org
PO Box 841
Eatontown, NJ 07724
ph: 732-531-9668
pegdiver@earthlink.net
Promotes a positive and safe image of diving and represents the diving community through its various committees. The council is composed of twenty-one member clubs plus individual members. Its environmental concerns include access to marine protected areas, beach replenishment, and antipollution work, including creation of a "Clean Ocean Zone."

New Jersey PIRG
www.njpirg.org
11 N. Willow St.
Trenton, NJ 08608
ph: 609-394-8155
info@njpirg.org
Educates citizens to participate in the political process, including fisheries reform and other marine issues.

New Jersey State Federation of Sportsmen
www.njsfsc.org
PO Box 751
Chester, NJ 07930
ph: 908-879-6321
fax: 908-879-6003
webmaster@njsfsc.org
Seeks to foster in the people of New Jersey a keen awareness of the many priceless riches upon its soils, waters,
forests, minerals, plants, wildlife, and natural geographic
features and to build and encourage in these people a
commitment to the wise use and proper management of
those resources for the benefit of all New Jersey, present
and future.

New York–New Jersey Baykeeper
www.nynjbaykeeper.org
52 West Front St.
Keyport, NJ 07735
ph: 732-888-9870
fax: 732-888-9873
info@nynjbaykeeper.org
Seeks to protect, preserve, and restore the ecological
integrity and productivity of the Hudson-Raritan Estuary.
Baykeeper serves as citizen advocate for the estuary's bays,
streams, and shores while stopping polluters, championing
public access, influencing land use decisions, and restoring
habitat benefiting the natural and human communities of
the watershed.

Save Barnegat Bay
www.savebarnegatbay.org
906-B Grand Central Ave.
Lavallette, NJ 08735
ph: 732-830-3600
fax: 732-830-6670
info@savebarnegatbay.org
Works to conserve undeveloped natural land and clean
water throughout the Barnegat Bay watershed. Founded in
1972, this not-for-profit environmental group is a local
chapter of the Izaak Walton League of America, a leading
national environmental organization.

Seas Monmouth
www.sailseas.com/monmouth
2099 Whittler St.
Rahway, NJ 07065
Promotes and encourages small-craft activity as recreation
and sport by providing low-cost learn-to-sail programs and
encourages relationships among all people interested in
sailing craft and water safety.

Shark Research Institute
www.sharks.org
PO Box 40
Princeton, NJ 08540
ph: 609-921-3522
fax: 609-921-1505
marie@sharks.org
Sponsors and conducts research on sharks and promotes
the conservation of sharks. SRI's headquarters are in
Princeton, New Jersey. It also has field offices in Canada,
the Galapagos Islands, Honduras, Mexico, South Africa, and
the Seychelles. Australia is home to a new data-collecting
site.

Shark River Cleanup Coalition Inc.
www.sharkriver.org
PO Box 2241
Neptune City, NJ 07753
ph: 732-775-7196
fax: 732-775-7196
Seeks to enhance the water quality of the Shark River
Estuary and its fresh water tributaries, to improve and protect habitats important to the conservation and abundance
of the wildlife, and to protect the recreational and commercial uses from degradation and pollution, thereby ensuring
the ecological and economical stability of the watershed.

Sierra Club–New Jersey Chapter
www.sierraactivists.org
139 West Hanover St.
Trenton, NJ 08618
ph: 609-656-7612
fax: 609-656-7618
njdirecter@njsierra.org
Uses lawful means to practice and promote the responsible
use of the earth's ecosystems and resources, and to educate and enlist humanity to protect and restore the quality
of the natural and human environment. Coastal chapters
within the state include Jersey Shore, North Jersey, South
Jersey, and Loantaka group. Work includes fighting a state
rule that would roll back thirty years of coastal conservation efforts.

Surfers' Environmental Alliance
www.seasurfer.org
1032 Woodgate Ave.
Long Branch, NJ 07740
ph: 917-453-7311
fitzpatrickunger@yahoo.com
Seeks to protect the natural wonders of the coastal environment and to foster and protect beach access. SEA is a
grassroots, project-based organization aggressively committed to the cultural and environmental integrity of the sport
of surfing.

Surfrider Foundation–Jersey Shore
www.surfrider.org/jerseyshore
PO Box 760
Belmar, NJ 07719-0760
jerseyshore@surfrider.org
Uses conservation, activism, research, and education to
encourage people to protect and enjoy the world's waves
and beaches. The Surfrider Foundation is a nonprofit environmental organization whose membership consists mainly
of surfers. Its scope has expanded to include protecting
wetlands, bird life, and beaches as well as surf spots. It has
sixty-six chapters worldwide, including two chapters in
New Jersey.

**Surfrider Foundation–South Jersey Organizing
Committee**
46 Northfield Ave.
Northfield, NJ 08225
ph: 609-485-0217

Tuckerton Seaport
www.Tuckertonseaport.org
120 West Main St.
PO Box 52
Tuckerton, NJ 08087
ph: 609-296-8868
fax: 609-296-5810
info@tuckertonseaport.org
Strives to preserve, present, and interpret the rich maritime
history, artistry, heritage, and environment of the Jersey
shore and the unique contributions of its baymen.

Waterspirit
www.waterspirit.org
981 Ocean Ave.
Elberon, NJ 07740
ph: 732-923-9788
fax: 732-229 -8960
unws@bellatlantic.net
Offers opportunities to become familiar with water issues
and to view them in the context of the sacredness and
interrelatedness of all creation.

NEW MEXICO
Acoustic Ecology Institute
www.acousticecology.org
45 Cougar Canyon
Santa Fe, NM 87508
ph: 505-466-1879
fax: 505-466-4930
info@eacousticecology.org
Works to increase personal and social awareness of the
sound environment and to contribute toward development
of ethical public policies regarding sounds on the land and
in the world's oceans where human-generated sounds from
military, industrial, and commercial operations are beginning to impact marine fish and mammals.

National Tribal Environmental Council
www.ntec.org
2501 Rio Grande Blvd. NW, Ste. A
Albuquerque, NM 87104
ph: 505-242-2175
fax: 505-242-2654
info@ntec.org
Works with most of the coastal tribes involved in marine
fisheries, cultural, and resource protection activities. The
council includes marine sessions at its national conference.

Planetary Coral Reef Foundation
www.pcrf.org
9 Silver Hills Rd.
Santa Fe, NM 87505
ph: 505-474-7444
fax: 505-424-3336
alling@pcrf.org
Works to preserve and protect the earth's coral reefs
through pioneering programs in science, technology, and
education. This includes providing coral reef field research
experience and seamanship training to an international
team of students aboard the 84-foot sailing research vessel
RV Heraclitus.

NEW YORK
American Littoral Society–Northeast Chapter
www.alsnyc.org
28th West 9th Rd.
Broad Channel, NY 11693
ph: 718-318-9344
driepe@nyc.rr.com
Draws attention to environmental issues that impact upon
the coastal zone.

American Museum of Natural History
http://research.amnh.org/biodiversity
Central Park West at 79th St.
New York, NY 10024
ph: 212-769-5742
fax: 212-769-5292
biodiversity@amnh.org
Mitigates critical threats to global biodiversity and maintains a large oceans exhibition hall.

Aquarium for Wildlife Conservation

www.nyaquarium.com
Surf Ave. at West 8th St.
Brooklyn, NY 11224
ph: 718-265-FISH
nyageneral@wcs.org
Saves wildlife and wild places around the globe. The aquarium was formerly known as the Coney Island Aquarium.

Atlantic Salmon Fish Creek Club

www.dreamscape.com/flyman
PO Box 67
Sylvan Beach, NY 13157
ph: 315-337-4079
pmiller3825@yahoo.com
Develops a supportable restoration effort for Atlantic salmon in the Oswego River system.

Atlantis Marine World Aquarium

www.atlantismarineworld.com
431 E. Main St.
Riverhead, NY 11901-2250
ph: 631-208-9200
Features marine life exhibits, sea lion training sessions, educational programs, and a submarine simulator ride. Located in Riverhead along the Peconic River, the aquarium is also home to the Riverhead Foundation for Marine Research and Preservation, New York State's only authorized stranding and rehabilitation facility.

Audubon Society–New York, Great South Bay

www.gsbas.com/
PO Box 267
Sayville, NY 11782
ph: 631-563-7716
Advocates for conservation of habitat for birds and other wildlife of Long Island. Audubon is a national network of community-based nature centers and chapters, scientific and educational programs, and advocacy on behalf of areas sustaining important bird populations.

Blue Ocean Institute

www.blueoceaninstitute.org
250 Lawrence Hill Rd.
Cold Spring Harbor, NY 11724
ph: 631-367-0063
fax: 631-367-7279
info@blueoceaninstitute.org
Works to inspire a closer relationship with the sea through science, art, and literature. The nonprofit institute also works to develop conservation solutions that are compassionate to people as well as to ocean wildlife.

Citizens Campaign for the Environment

www.citizenscampaign.org
225A Main St.
Farmingdale, NY 11735
ph: 516-390-7150
fax: 516-390-7160
Works to build widespread citizen understanding and advocacy for policies and actions designed to manage and protect interdependent land and water resources, wildlife, and public health. CCE carries out this mission through public education, research, lobbying, organizing, and public outreach.

Clamshell Foundation

www.ehsandcastle.com
PO Box 2725
East Hampton, NY 11937
ph: 631-324-6250
fax: 631-329-5977
info@ehsandcastle.com
Raises money to benefit community and environmental programs.

Coalition to Save Hempstead Harbor

www.mysound.uconn.edu/hemphbr_org.html
247 Sea Cliff Ave.
PO Box 159
Sea Cliff, NY 11579
ph: 516-759-3832
fax: 516-759-1156
cshh@earthlink.net
Seeks to identify and eliminate environmental threats to Hempstead Harbor and surrounding communities and to advance the public interest in restoring and protecting the local environment to its full ecological and economic potential.

Concerned Citizens of Montauk

PO Box 146
Montauk, NY 11954
montaukian@aol.com
Works to protect the Montauk area's natural resources and environment.

Crescent Beach Civic Association

180 Wiman Ave.
Staten Island, NY 10308
ph: 718-984-9840
fax: 718-967-3969
gkharborpark@aol.com
Strives to protect a popular family recreational beach from development.

East Hampton Baymen's Association
130 Gerard St.
East Hampton, NY 11937
Works to maintain sustainable fishing livelihoods and culture on Long Island.

Environmental Background Information Center
www.ebic.org
45 West 21st St., 4th Fl.
New York, NY 10007
Provides environmental background information on corporate polluters to grassroots organizations fighting to protect Long Island Sound and for other causes.

Environmental Defense–New York
www.environmentaldefense.org/home.cfm
257 Park Ave. S
New York, NY 10010
ph: 212-505-2100
fax: 212-505-2375
press@environmentaldefense.org
Finds practical solutions to environmental problems, including threats to the marine environment, with particular focus on fisheries and marine protected areas.

Fire Island Ecology
www.firei.org
fire_island_ecology@earthlink.net
Provides information about Fire Island that focuses on the environment and the remarkable natural values of the barrier island.

Fisheries Defense Fund Inc.
510 Park Ave., #8B
New York, NY 10022-1105
ph: 212-688-7567
fax: 212-688-7567
fishsave@pipeline.com
Protects fishery resources through advocacy and education.

Fish Unlimited
www.fishunlimited.org
PO Box 1073
Shelter Island, NY 11965
ph: 631-749-3474
fax: 631-749-3476
info@fishunlimited.org
Advocates responsible stewardship of fresh water and saltwater estuaries.

Friends of Fire Island
www.ffins.org
PO Box 504
Patchogue, NY 11772-0504
Provides support for Fire Island's National Seashore independent of the National Park Service.

Friends of the Bay
www.friendsofthebay.org
PO Box 564
Oyster Bay, NY 11771
ph: 516-922-6666
fax: 516-385-9817
info@friendsofthebay.org
Represents nearly 3,500 Oyster Bay-area residents and businesses and is one of the most effective environmental organizations around Long Island Sound. Friends' work includes water quality monitoring, wetlands restoration, and active engagement in land-use planning around the bay.

Group for the South Fork
http://thehamptons.com/group/main.html
PO Box 569
2442 Main St.
Bridgehampton, NY 11932
ph: 631-537-1400
fax: 631-537-2201
staff@thehamptons.com
Seeks to protect the environment, rural character, and quality of life across the South Fork through public advocacy and public education.

Hudson River Foundation
www.hudsonriver.org
17 Battery Place, Ste. 915
New York, NY 10004
ph: 212-HUDSONR
fax: 212-924-8325
info@hudsonriver.org
Makes science integral to decision making with regard to the Hudson River from its headwaters to the sea.

Hudson Riverkeeper
www.riverkeeper.org
PO Box 130
Garrison, NY 10524
ph: 800-21-RIVER
info@riverkeeper.org
Safeguards the integrity of the Hudson River, its tributaries, and the watershed of New York City.

Hudson River Sloop *Clearwater*

www.clearwater.org
112 Little Market St.
Poughkeepsie, NY 12601
ph: 845-454-7673
office@clearwater.org
Created to defend and restore the Hudson River, this classic river sloop has become a magnet for public attention and activism around river restoration.

The Metropolitan Waterfront Alliance

www.Waterwire.net
457 Madison Ave.
New York, NY 10022
ph: 800-364-9943
fax: 888-486-9688
info@waterwire.net
Works to help the region reclaim and reconnect to its greatest natural resource—the harbor, rivers, and estuaries of the New York and New Jersey waterfront.

Montauk Boatmen's and Captains' Association

www.mymatecharterboat.com
PO Box 1908
East Hampton, NY 11937
ph: 631-329-0973
fax: 631-329-6560
captainmcbride@optonline.net
Comprises the largest group of professional charter party boat captains sailing out of the same port. The association's concerns include maintaining a sustainable resource offshore.

National Audubon Society

www.audubon.org
700 Broadway
New York, NY 10003
ph: 212-979-3000
fax: 212-979-3188
audubonaction@audubon.org
Seeks to conserve and restore natural ecosystems while focusing on birds, other wildlife, and their habitats for the benefit of humanity and the earth's biological diversity. Audubon is a national network of community-based nature centers and chapters, scientific and educational programs, and advocacy on behalf of areas sustaining important bird populations. As part of its marine work, the society also produces a guide to sustainable seafood.

Natural Resources Defense Council–New York

www.nrdc.org
40 W. 20th St.
New York, NY 10011
ph: 212-727-2700
fax: 212-727-1773
nrdcinfo@nrdc.org
Uses law, science, and the support of more than one million online activists to protect the planet's wildlife and wild places and to ensure a safe and healthy environment for all living things. NRDC also has an active marine program.

Natural Resources Protective Association

www.nrpa.com
PO Box 050328
Staten Island, NY 10305
ph: 718-987-6037
kerrysull@si.rr.com
Works to protect the marine environment of Raritan Bay and Lower New York Harbor.

The Nature Conservancy–New York City

http://nature.org/wherewework/northamerica/states/
newyork
570 Seventh Ave., Ste. 601
New York, NY 10018
ph: 212-381-2181
fax: 212-997-8451
ralovera@tnc.org
sdennin@tnc.org
Seeks to preserve the plants, animals, and natural communities that represent the diversity of life on earth through land purchases, establishing conservation easements, and other means. The Nature Conservancy is a national organization headquartered in Arlington, Virginia. In New York TNC is working to protect a range of habitats, including the Peconic estuary, Long Island pine barrens, ocean beaches and bays, and the Montauk Peninsula.

New York Coalition for Alternatives to Pesticides

353 Hamilton St.
Albany, NY 12210
nycap@crisny.org
Fights to replace a major source of coastal pollution—nutrient-heavy synthetic pesticides.

New York League of Conservation Voters

www.nylcv.org
29 Broadway, Ste. 1100
New York, NY 10006
ph: 212-361-6350
fax: 212-361-6363
info@nylcv.org
Mobilizes New Yorkers as a political force on behalf of the environment, including protection of the state's coastal and marine resources.

New York PIRG
www.nypirg.com
9 Murray St.
New York, NY 10007
ph: 212-349-6460
fax: 212-349-1366
nyc@nypirg.org
Educates citizens to participate in the political process,
including fisheries reform and other marine issues.

The New York Whale and Dolphin Action League
www.ny4whales.org
PO Box 273
Tuckahoe, NY 10707
ph: 914-793-9186
ny4whales@optonline.net
Seeks legislation that will preserve oceans, coastal regions,
waterways, and marine life. The league is a project of
Cetacean Society International, which is dedicated to the
preservation of marine life, and especially cetaceans
(whales and dolphins).

North Fork Environmental Council
www.nfec1.org
PO Box 799
Mattituck, NY 11952
ph: 631-298-8880
fax: 631-298-4649
nfec@optonline.net
Works to preserve land, sea, and air on Long Island's
North Fork.

**North Shore Waterfront Conservancy of
Staten Island**
www.nswcsi.org
PO Box 140502
Staten Island, NY 10314
ph: 718-273-2329
nswcsi@yahoo.com
Provides education on environmental and related community
health issues, and advocates for economic development as
it relates to the environment. The conservancy focuses on
the north shore along the Kill van Kull from St. George to
Arlington and is concerned with community-based planning.

**Pace Environmental Litigation Clinic
Pace University School of Law**
www.law.pace.edu/envclinic
78 N. Broadway, Bldg. E
White Plains, NY 10603
ph: 914-422-4343
fax: 914-422-4437
mpostman@law.pace.edu

Immerses students in environmental law practice by representing public interest groups such as Waterkeepers, which
fight to stop pollution of rivers, bays, and coasts.

Peconic Baykeeper
www.peconicbaykeeper.org
PO Box 1308
206 Lincoln St.
Riverhead, NY 11901
ph: 631-727-7346
fax: 631-727-9518
info@peconicbaykeeper.org
Works to protect and sustain the Peconic and South Shore
estuaries on Long Island through improving water quality
and productivity by emphasizing the intrinsic connection
between bays and communities.

Pew Institute for Ocean Science
www.pewocean.org
126 East 56th St.
New York, NY 10022
ph: 212-756-0042
fax: 212-756-0045
Promotes world-class scientific activity aimed at protecting
the world's oceans and the species that inhabit them.

**Riverhead Foundation for Marine Research and
Preservation**
www.riverheadfoundation.org
467 East Main St.
Riverhead, NY 11901
ph: 631-369-9840
fax: 631-369-9826
dwilliams@riverheadfoundation.org
Researches projects that examine the biology and ecology of
marine environments and rescues and rehabilitates stranded
marine animals. (See also Atlantis Marine World Aquarium.)

Save Our Seashore
www.saveourseashore.org
PO Box 469
Brightwaters, NY 11718-0469
ph: 631-583-9266
fax: 631-583-9268
sosny@aol.com
Works to preserve Long Island's shoreline.

Save the Beaches Fund Inc.
www.savethebeaches.org
PO Box 622
Babylon, NY 11702-0622
ph: 631-786-8060
fax: 631-786-7042
stbfinc@aol.com
Preserves and protects coastal heritage, wildlife, and
wetlands.

Save the Oceans
www.Save-the-oceans.org
P.O. Box 212
Hastings-on-Hudson, NY 10706-0212
ph: 914-478-2726
fax: 914-478-2667
info@oceansalert.org
Helps to restore the health of the oceans and waterways
and to rebuild healthy and sustainable fish populations and
habitats through forums and as a coordinator of international meetings using "people power" to achieve the aims
of the mission. The organization works for healthy fish
through regulation and certification programs. It works to
reduce pollutants, and uses mass media to create awareness of the state of the oceans.

Save the Sound Inc.–New York Regional Office
www.savethesound.org
c/o Garvies Pt. Museum
50 Barry Dr.
Glen Cove, NY 11542
ph: 888-SAVE-LIS
savethesound@savethesound.org
Seeks restoration, protection, and appreciation of Long
Island Sound and its watershed through advocacy, education, and research.

Scenic Hudson
www.scenichudson.org
One Civic Center Plaza, Ste. 200
Poughkeepsie, NY 12601
ph: 845-473-4440
fax: 845-473-0740
ablcklng@scenichudson.org
Seeks to ensure that the Hudson River Valley remains environmentally and economically sustainable.

Seatuck Environmental Association
www.seatuck.org
PO Box 31
Islip, NY 11751
ph: 631-581-6908
staff@seatuck.org
Promotes the conservation of Long Island's wildlife and
environment through education, research, and advocacy.

Sierra Club–New York City Group
http://newyork.sierraclub.org
40 Exchange Pl., Ste. 2010
New York, NY 10005
ph: 646-405-9838
Promotes conservation of the natural environment through
public education and lobbying, including protection of
urban wetlands and water quality along the Hudson River.

Staten Island Friends of Clearwater
400 Delaware Ave.
Staten Island, NY 10303
ph: 718-987-6037
fax: 718-987-6037
Works to protect wildlife, waterways, the shoreline, and
public access to natural areas. The group also organizes
beach cleanups and habitat restoration projects.

**Surfrider Foundation–New York,
Long Island Chapter**
www.surfrider.org/longisland
PO Box 2681
Amagansett, NY 11930
ph: 631-329-4012
eugene@surfriderli.org
Uses conservation, activism, research, and education to
encourage people to protect and enjoy the world's waves
and beaches. The Surfrider Foundation is a nonprofit environmental organization whose membership consists mainly
of surfers. Its scope has expanded to include protecting
wetlands, bird life, and beaches as well as surf spots. It has
sixty-six chapters worldwide, including two chapters in
New York.

**Surfrider Foundation–New York, New York City
Chapter**
www.surfrider.org/nyc
PO Box 257
New York, NY 10014-0257
ph: 646-515-9290
surfridernyc@surfrider.org

U.S. Merchant Marine Academy
www.usmma.edu
300 Steamboat Rd.
Kings Point, NY 11024
ph: 516-773-5888
Ensures that talented and well-trained people are available
to the nation as shipboard officers and as leaders in the
transportation field. Also the training site for NOAA Corps,
the uniformed service of the nation's leading civilian ocean
agency.

Waterkeeper Alliance
www.waterkeeper.org
828 S. Broadway, Ste. 100
Tarrytown, NY 10591
ph: 914-674-0622
info@waterkeeper.org
Works to protect public waters from polluters. This is the
national headquarters of the more than 100 waterkeepers.

Wildlife Conservation Society/Bronx Zoo
http://wcs.org
2300 Southern Blvd.
Bronx, NY 10460
ph: 718-220-5100
fax: 718-220-2685
epikitch@wcs.org
Works to save wildlife and wild lands through careful science, conservation, and education. This organization is also known as the Bronx Zoo.

NORTH CAROLINA
Cape Fear River Watch, Inc.
http://cfrw.us/
617 Surry St.
Wilmington, NC 28401
ph: 910-762-5606
cfrw@ecoisp.com
Seeks to protect and improve the water quality of the Lower Cape Fear River Basin through education, advocacy, and action.

Cape Hatteras Coastkeeper
www.nccoast.org/ch-coastkeeper.htm
PO Box 475
Manteo, NC 27954
ph: 252-473-1607
hatteraskeeper@nccoast.org
Monitors the tidal creeks and rivers, sounds, barrier islands, and nearshore waters along the northern coast from the southern end of Ocracoke Island to the Virginia state line and inland.

Cape Lookout Coastkeeper
www.nccoast.org/ch-coastkeeper.htm
3609 Highway 24 (Ocean)
Newport, NC 28570
ph: 252-393-8185
lookoutkeeper@nccoast.org
Monitors the tidal creeks and rivers, sounds, barrier islands, and nearshore waters from the New River inlet near Jacksonville to the southern end of Ocracoke Island, covering about 1,200 square miles of water.

Conservation Council of North Carolina
www.conservationcouncilnc.org
PO Box 12671
Raleigh, NC 27605
ph: 919-839-0006
ccnc@conservationcouncilnc.org
Lobbies on behalf of North Carolina's environment, including coastal and marine protection.

Duke University Marine Laboratory
www.nicholas.duke.edu/marinelab/
135 Duke Marine Lab Rd.
Beaufort, NC 28516-9721
ph: 252-504-7503
fax: 252-504-7648
Focuses on marine policy and conservation science.

Environmental Defense–Southeast Atlantic Office
2500 Blue Ridge Rd., Ste. 330
Raleigh, NC 27607
ph: 919-881-2601
Works to develop sustainable harvest plans for marine and estuarine fishes; to protect the major estuaries of the region from water quality degradation; to safeguard critical fish spawning areas and nurseries; to sustain coral reefs, seagrass beds, shellfish beds and other essential habitats; and to combine these approaches into one of the first true ecosystem-based management regimes anywhere.

Lower Neuse Riverkeeper
www.neuseriver.org/RiverKeeper.html
220 S. Front St.
PO Box 15451
New Bern, NC 28561
ph: 252-637-1970
fax: 252-637-1970
riverkeeper@neuseriver.org
Seek to locate and eliminate pollution sources on the Neuse River, tributaries, and shorelines by use of boats, aircraft, and trucks.

The Nature Conservancy–North Carolina Chapter
http://nature.org/wherewework/northamerica/states/
northcarolina/
One University Place, Ste. 290
4705 U Dr.
Durham, NC 27707
ph: 919-403-8558
fax: 919-403-0379
ncchapter@tnc.org
Works to protect over 105 sites across the state, including on the Outer Banks and other coastal areas. The Nature Conservancy is a national organization headquartered in Arlington, Virginia. Through land purchases, establishing conservation easements and other means, it seeks to preserve the plants, animals, and natural communities that represent the diversity of life on earth.

North Carolina Aquariums

www.ncaquariums.com
417 North Blount St.
Raleigh, NC 27601
ph: 800-832-FISH
fax: 919-733-471
admin@ncaquariums.com
Promote an awareness, understanding, appreciation, and conservation of the diverse natural and cultural resources associated with North Carolina's aquatic environments. The aquariums were established in 1976.

North Carolina Aquarium–Fort Fisher

www.ncaquariums.com
900 Loggerhead Rd.
Kure Beach, NC 28449
ph: 866-301-3476
fax: 910-458-6812
ffmail@ncaquariums.com

North Carolina Aquarium–Pine Knoll Shores

www.ncaquariums.com
1010 W. Fort Macon Rd., Ste. 18B
Atlantic Beach, NC 28512
ph: 866-294-3477
fax: 252-247-0663
pksmail@ncaquariums.com
This aquarium is presently closed for expansion and will reopen in 2006.

North Carolina Aquarium–Roanoke Island

www.ncaquariums.com
PO Box 967
Manteo, NC 27954-0967
ph: 888-332-3475
rimail@ncaquariums.com

North Carolina Coastal Federation

www.nccoast.org
3609 Highway 24 (Ocean)
Newport, NC 28570
ph: 252-393-8185
fax: 252-393-7508
nccf@nccoast.org
Seeks to aid citizens and local groups to take an active role in the wise management of North Carolina's coastal water quality and natural resources. Founded in 1982 by eight grassroots groups concerned with coastal issues, NCCF today serves 7,500 members and 200 affiliated organizations.

North Carolina PIRG

www.ncpirg.org
112 South Blount St.
Raleigh, NC 27601
ph: 919-833-2070
fax: 919-839-0767
info@ncpirg.org
Educates citizens to participate in the political process, including fisheries reforms and other marine issues.

North Carolina Wildlife Federation

www.ncwf.org
1204 The Plaza, #2
Charlotte, NC 28205
ph: 704-377-4696
ncwf_charlotte@mindspring.com
Strives to be the leading advocate for all North Carolina wildlife and its habitats. Their aquatic habitat campaign seeks to protect North Carolina's fields, shorelines, and coastlines.

Pamlico-Tar River Foundation

www.ptrf.org
PO Box 1854
Washington, NC 27889
ph: 252-946-7211
fax: 252-946-9492
info@ptrf.org
Works to protect, preserve, and promote the environmental quality of the Tar-Pamlico River and its watershed.

Sierra Club–North Carolina Chapter

www.sierraclub.org/nc/
112 S. Blount St.
Raleigh, NC 27601
ph: 919-833-8467
fax: 919-833-8460
info@sierraclub-nc.org
Uses lawful means to practice and promote the responsible use of the earth's ecosystems and resources, and to educate and enlist humanity to protect and restore the quality of the natural and human environment. The Sierra Club is America's oldest, largest, and most influential grassroots environmental group. The North Carolina chapter works to protect areas such as North Carolina's Outer Banks, one of the few remaining natural coastal barrier island systems in the world.

Southern Environmental Law Center

www.southernenvironment.org
200 West Franklin St., Ste. 330
Chapel Hill, NC 27516
ph: 919-967-1450
fax: 919-929-9421
Strengthens environmental laws and policies throughout the south. Works with partners to oppose a planned navy airport that threatens coastal areas used by migratory birds

Surfrider Foundation–North Carolina, Outer Banks Chapter
www.surfrider.org/outerbanks
PO Box 1576
Kill Devil Hills, NC 27948
ph: 252-491-8639
rednsis@aol.com
Uses conservation, activism, research, and education to encourage people to protect and enjoy the world's waves and beaches. The Surfrider Foundation is a nonprofit environmental organization whose membership consists mainly of surfers. Its scope has expanded to include protecting wetlands, bird life, and beaches as well as surf spots. It has sixty-six chapters worldwide, including two chapters in North Carolina.

Surfrider Foundation–North Carolina, Topsail Chapter
www.surfrider.org/topsail
PO Box 4068
Surf City, Topsail Island NC 28445
ph: 910-328-3147

OREGON

Audubon Society–Oregon, Cape Arago
PO Box 381
North Bend, OR 97459
topits@gte.net
Works to conserve and restore natural ecosystems, focusing on birds, other wildlife, and their habitats for the benefit of humanity and the earth's biological diversity. This chapter works to protect Pacific coastal habitat in areas near Coos Bay.

Audubon Society of Portland
www.audubonportland.org
5151 NW Cornell Rd.
Portland, OR 97210
ph: 503-292-6855
fax: 503-292-1021
general@audubonportland.org

Coast Range Association
www.coastrange.org
PO Box 2250
Corvallis, OR 97339
ph: 541-752-3146
information@coastrange.org
Works toward an environmentally sustainable economy in Oregon.

CoastWatch
www.oregonshores.org
605 SE 37th Ave.
Portland, OR 97214
ph: 503-238-4450
orshores@teleport.com
Works to monitor and protect the Oregon coast. Volunteers adopt a mile of coast.

COMPASS—Communications Partnership for Science and the Sea
www.compassonline.org
COMPASS Program Manager
Oregon State University
Department of Zoology
3029 Cordley Hall
Corvallis, OR 97331
ph: 541-737-9173
Strives to accelerate the pace of solutions to important marine environmental problems. COMPASS also acts as a kind of public relations office for marine scientists reporting on critical new research relating to the health of the oceans.

Conservation Biology Institute
www.consbio.org
260 SW Madison Ave., Ste. 106
Corvallis, OR 97333
ph: 541-757-0687
fax: 541-757-0518
stritt@consbio.org
Uses applied conservation research and education to help save the diversity of life on earth.

Corvallis Environmental Center
www.peak.org/~ecenter
214 SW Monroe Ave.
PO Box 2189
Corvallis, OR 97339
ph: 541-753-9211
ecenter@peak.org
Provides opportunities for Corvallis community members to learn about their connection to the natural environment, including Oregon's coasts and ocean.

EcoTrust
www.ecotrust.org
721 NW Ninth Ave., Ste. 200
Portland, OR 97209
ph: 503-227-6225
fax: 503-227-1517
info@ecotrust.org
Seeks to build a place where people and wild salmon thrive.

Hatfield Marine Science Center
http://hmsc.oregonstate.edu/visitor
2030 S. Marine Science Dr.
Newport, OR 97365
ph: 541-867-0100
hmsc.@oregonstate.edu
Offers a unique and dynamic environment for lifelong
marine exploration and discovery.

North Coast Watershed Association
www.clatsopwatersheds.org/
750 Commercial St., Rm. 205
Astoria, OR 97103
ph: 503-325-0435
fax: 503-325-0459
tcullison@columbiaestuary.org
An umbrella of four watershed councils working to restore
salmon and watersheds in the Columbia River Estuary and
the north coast of Clatsop County, Oregon. Offers a link
between local governments, community groups, businesses,
and local residents.

Northwest Environmental Advocates
www.northwestenvironmentaladvocates.org
PO Box 12187
Portland, OR 97212-0187
ph: 503-295-0490
nbell@advocates-nwea.org
Uses a mix of approaches—including negotiation, litigation, education, community organizing, and advocacy—to
restore water and wetlands, air quality, and wildlife and
develop renewable energy in the Northwest.

Ocean Defense International
www.oceandefense.org
PO Box 3464
Ashland, OR 97520
Combats the destruction of the oceans and its inhabitants.

1000 Friends of Oregon
www.friends.org
534 Southwest Third Ave., Ste. 300
Portland, OR 97204
ph: 503-497-1000
fax: 503-223-0073
info@friends.org
Works to protect Oregon's quality of life through the conservation of farm and forest lands, protection of natural
and historic resources, and promotion of more compact and
livable cities. The Oregon Coastal Futures Project is one of
the "Friends" major focuses through 2006.

Oregon Coast Aquarium
www.aquarium.org
2820 SE Ferry Slip Rd.
Newport, OR 97341
ph: 541-867-3474
fax: 541-867-3474
webmaster@aquarium.org
Offers a view into the past, present, and future of ocean
exploration through its marine exhibits and educational
programs. The aquarium was formerly the home of Keiko,
the famed orca.

Oregon Environmental Council
www.orcouncil.org
222 NW Davis St., Ste. 309
Portland, OR 97209
ph: 503-222-1963
fax: 503-222-1405
info@oeconline.org
Strives to bring Oregonians together for a healthy environment. Founded in 1968, the Oregon Environmental Council
is a nonprofit, nonpartisan organization with more than
2,000 members who work on a range of issues, including
the creation of coastal watershed councils.

Oregon Natural Resources Council
www.onrc.org
5825 N. Greeley Ave.
Portland, OR 97217
info@onrc.org
Protects and restores Oregon's wild lands, wildlife, and
waters.

Oregon Shores Conservation Coalition
www.oregonshores.org
16005 NW Rockyford Rd.
Yamhill, OR 97148
ph: 503-662-4210
fromherz@easystreet.com
Works to protect Oregon's coastal environment and the
public trust therein through education, advocacy, and
engaging citizens to keep watch over and defend the
Oregon coast.

Oregon State PIRG
www.ospirg.org
1536 SE 11th Ave.
Portland, OR 97214
ph: 503-231-4181
fax: 503-231-4007
info@ospirg.org
Educates citizens to participate in the political process,
including fisheries reform and other marine issues.

Oregon Trout

www.oregontrout.org
117 SW Naito Parkway
Portland, OR 97204
ph: 503-222-9091
info@ortrout.org
Protects and restores native wild fish and their habitats.

Oregon Wildlife Federation

www.orwildlife.org
PO Box 5878
Portland, OR 97228-5878
ph: 503-234-2694
blc@hevanet.com
Links and supports grassroots activists throughout Oregon
who are working to protect the state's rivers, ancient
forests, deserts, oceans, and the species that depend upon
them for survival. OWF is one of Oregon's oldest conservation organizations.

Pacific Coast Federation of Fishermen's Associations (PCFFA) Northwest Office/The Institute for Fisheries Resources

www.ifrfish.org
PO Box 11170
Eugene, OR 97440-3370
ph: 541-689-2000
fax: 541-689-2500
fish1ifr@aol.com
Works to ensure the rights of individual fishermen and
fights for the long-term survival of commercial fishing as a
productive livelihood and way of life. The federation is the
largest and most politically active trade association of commercial fishermen on the West Coast.

Pacific Marine Conservation Council–Oregon Office

www.pmcc.org
PO Box 59
Astoria, OR 97103
ph: 503-325-8188
fax: 503-325-9681
deborah@pmcc.org
Represents fishing communities and concerned citizens
dedicated to sustaining healthy and diverse marine ecosystems on the West Coast.

Pacific Rivers Council

www.pacrivers.org
PO Box 10798
Eugene, OR 97440
ph: 541-345-0119
fax: 541-345-0710
info@pacrivers.org
Works to protect and restore rivers, their watersheds, and
native aquatic species.

Port Orford Ocean Resources Team (POORT)

PO Box 679
Port Orford, OR 97465
ph: 541-332-0627
poort@carrollsweb.com
Supports a community-based planning project to help create local fishery management decisions that are sustainable
for the fish population and the economic viability of the
Port Orford community.

River Network

www.rivernetwork.org
520 SW 6th Ave., #1130
Portland, OR 97204
ph: 503-241-3506
fax: 503-241-9256
info@rivernetwork.org
Strives to help people understand, protect, and restore
rivers and their watersheds.

Salmon for All

www.salmonforall.org
PO Box 56
Astoria, OR 97103
ph: 503-325-3831
fax: 503-325-2725
salforal@pacifier.com
Provides ongoing education concerning the public harvest
industry, takes active advocacy roles in legislative and fisheries deliberations, and works to ensure the health of the
Columbia River and supports its responsible use among
Treaty Native nations, industry, and recreational users.

Sierra Club–Oregon, Columbia

www.oregon.sierraclub.org/groups/columbia/index.asp
2950 SE Stark, Ste. 100
Portland, OR 97214
ph: 503- 238-0442 x300
Promotes conservation of the Oregon natural environment
by influencing public policy decisions—legislative, administrative, legal, and electoral—on a range of issues, including
coastal and ocean protection. The Sierra Club is America's
oldest, largest, and most influential grassroots environmental group.

Sierra Club–Oregon, Marys Peak

http://oregon.sierraclub.org/groups.maryspeak/
PO Box 863
Corvalis, OR 97330
ph: 541-929-6272
wulff@peak.org

Sierra Club–Oregon Chapter
www.oregon.sierraclub.org
2950 SE Stark, Ste. 110
Portland, OR 97214
ph: 503-238-0442
fax: 503-238-6281
oregon.chapter@sierraclub.org

Surfrider Foundation–Central Oregon
www.surfrider.org/oregon
16 NW Kansas Ave.
Bend, OR 97701
ph: 541-317-5778
Uses conservation, activism, research, and education to
encourage people to protect and enjoy the world's waves
and beaches. The Surfrider Foundation is a nonprofit environmental organization whose membership consists mainly
of surfers. Its scope has expanded to include protecting
wetlands, bird life, and beaches as well as surf spots. It
has sixty-six chapters worldwide, including four chapters
in Oregon.

Surfrider Foundation–Pacific City
www.surfrider.org/oregon
PO Box 722
Pacific City, OR 97135
ph: 503-965-7873

Surfrider Foundation–Portland
www.surfrider.org/oregon
4236 SE Salmon St., Ste. D
Portland, OR 97215
rideoutas@hotmail.com

Surfrider Foundation–South Coast
www.surfrider.org/oregon
91382 Grinnell Ln.
Coos Bay, OR 97420
ph: 541-888-0710

West Coast Fishermen's Alliance
PO Box 5508
Charleston, OR 97420
ph: 541-888-3811
fax: 541-888-3811
Promotes sustainable fisheries, conservation, and the viability of small coastal communities. WCFA has approximately
sixty members from Washington to southern California, all
of whom fish commercially, mainly from small vessels.
Members longline for groundfish and target crab, salmon
and tuna.

Wild Salmon Center
www.wildsalmoncenter.org
721 NW 9th Ave., Ste. 290
Portland, OR 97209

ph: 503-222-1804
fax: 503-222-1805
info@wildsalmoncenter.org
Works in partnership with universities, governments and
private organizations from Russia, Canada, Japan, and the
United States to ensure a sustainable future for salmon
across the Pacific Rim.

PENNSYLVANIA
Delaware Riverkeeper Network
www.delawareriverkeeper.org
PO Box 326
Washington's Crossing, PA 18977-0326
ph: 215-369-1188
fax: 215-369-1181
drkn@delawareriverkeeper.org
Strengthens citizens protection of the Delaware River and
its tributary watersheds.

Independence Seaport Museum
www.phillyseaport.org
211 S. Columbus Blvd. & Walnut St.
Philadelphia, PA 19106
ph: 215-413-8615
fax: 215-925-8409
Captures the region's maritime heritage and connection to
the sea with family-oriented interactive exhibits, ship models, artifacts, and art. The museum also has a workshop on
the water dedicated to wooden boat building and educational programs, including marine conservation lessons
for children.

Pennsylvania Environmental Council
www.pecpa.org
130 Locust St., Ste. 200
Harrisburg, PA 17101
ph: 717-230-8044
fax: 717-230-8045
amcelwaine@pecpa.org
Plays an active role in environmental policy discussions and
decision making in Harrisburg (the state capitol), in both
the regulatory and legislative arenas. Seeks to protect
Pennsylvania's watersheds and find innovative solutions to
environmental problems.

Pennsylvania PIRG
www.pennpirg.org
1334 Walnut St., 6th Fl.
Philadelphia, PA 19107
ph: 215-732-3747
info@pennpirg.org
Educates citizens to participate in the political process,
including fisheries reform and other marine issues.

Pittsburgh Zoo and PPG Aquarium
www.pittsburghzoo.com
One Wild Place
Pittsburgh, PA 15206
ph: 800-474-4966; 412-665-3640
Features more than four thousand aquatic animals from
around the world and offers educational programs to help
visitors further explore the "mysteries of the deep."

PUERTO RICO
CORALations
www.coralations.org
PO Box 750
Culebra, PR 00775
ph: 877-77CORAL
fax: 530-618-4605
maryann@coralations.org
Works to maximize limited conservation resources by unit-
ing government agencies, academia, and the private sector
to accomplish specific coral reef conservation objectives.
CORALations conducts conservation projects using a multidisciplinary task force of scientists, engineers, and other
professionals as advisors.

La Asociación de Pescadores de Culebra
PO Box 240
Isla de Culebra, PR 00775
ph: 787-742-3371
corals@prtc.net (subj: Asociación la Pesca)
Maintains a marine-protected area to improve this commercial fishing group's livelihood and the ocean habitat that
ensures it.

Puerto Rico Coastkeeper/GuardAguas
PO Box 906-5112
San Juan, PR 00906-5112
ph: 787-723-4136
surfrider@caribe.net
Strives to maintain the health and integrity of the coast and
the ocean.

Rincon Organizing Committee
www.surfrider.org/rincon/
PO Box 1833
Rincon, PR 00677
ph: 787-823-2784
salvatrespalmas@yahoo.com
Seeks permanent protection against massive pending
development for the coastal area. The Surfrider Foundation
is a nonprofit environmental organization whose membership consists mainly of surfers. Its scope has expanded to
include protecting wetlands, bird life, and beaches as well
as surf spots. It has sixty-six chapters worldwide.

Vieques Waterkeeper
Soldado Serrano 52
Ocean Park, PR 00911
ph: 787-268-1300
Works to reclaim the island's land and waters formerly used
as a U.S. Navy bombing range. Waterkeeper Alliance is a
grassroots organization with local Waterkeeper programs
nationally and internationally dedicated to preserving and
protecting local waters from pollution.

RHODE ISLAND
Audubon Society–Rhode Island
www.asri.org
12 Sanderson Rd.
Smithfield, RI 02917
ph: 401-949-5454
fax: 401-949-5788
audubon@asri.org
Works in many venues of environmental policy including
water policy and protection of bay animals. Audubon's mission is to conserve and restore natural ecosystems, focusing
on birds, other wildlife, and their habitats for the benefit of
humanity and the earth's biological diversity.

Clean Water Action–Rhode Island
www.cleanwater.org
741 Westminster St.
Providence, RI 02903
ph: 401-331-6972
fax: 401-331-7072
provcwa@cleanwater.org
Works for clean, safe, affordable water.

Environmental Council of Rhode Island
www.environmentcouncilri.org/
PO Box 9061
Providence, RI 02940
ph: 401-621-8048
fax: 401-331-5266
environmentcouncil@earthlink.org
Serves as an effective voice for developing and advocating
policies and laws that protect and enhance the marine and
terrestrial environment of "The Ocean State."

Herreshoff Marine Museum
www.herreshoff.org
PO Box 450
One Burnside St.
Bristol, RI 02809
ph: 401-253-5000
fax: 401-253-6222
j.palmieri@herreshoff.org
Commemorates the Herreshoff Corporation and its contribution to boating and boating culture.

Metcalf Institute for Marine and Environmental Reporting
www.metcalfinstitute.org
Narragansett, RI 02882
ph: 401-874-6211
fax: 401-874-6486
jack@gso.uri.edu
Promotes clear and accurate reporting of scientific news and environmental issues and promotes the "blue beat," seeking to expand coverage of the other 71 percent of the planet that is saltwater.

Narrow River Preservation Association
www.narrowriver.org
PO Box 8
Sarnderstown, RI 02874
ph: 401-783-6277
nrpa@netsense.net
Preserves the quality of communities and the natural environment within the Narrow River watershed.

The Nature Conservancy–Global Marine Initiative
http://nature.org/marine
University of Rhode Island
Narragansett Bay Campus
South Ferry Rd.
Narragansett, RI 02882-1197
marine@tnc.org
Develops innovative strategies to protect the rich array of plant and animal life and safeguard the tremendous benefits the oceans provide. The Nature Conservancy is a national organization headquartered in Arlington, Virginia. Through land purchases, establishing conservation easements and other means, it seeks to preserve the plants, animals, and natural communities that represent the diversity of life on earth.

The Nature Conservancy–Rhode Island Chapter
http://nature.org/wherewework/northamerica/states/rhodeisland
159 Waterman St.
Providence, RI 02906
ph: 401-331-7110
fax: 401-273-4902
ri@tnc.org
Preserves the white pine forests, rivers, wetlands, and other natural systems that make Rhode Island unique.

Newport County Saltwater Fishing Club
PO Box 2
Newport, RI 02840
ph: 401-683-2812
fax: 401-683-2812
Promotes the protection and stewardship of fish and marine natural resources.

Newport's Friends of the Waterfront
www.newportwaterfront.org
PO Box 932
Newport, RI 02840
info@newportwaterfront.org
Protects public access to the water and fosters the sustainable development of the Newport waterfront.

The Ocean Project
www.theoceanproject.org
PO Box 2506
Providence, RI 02906
ph: 401-709-4071
fax: 401-272-8877
bmott@theoceanproject.org
Works to create an understanding among members of the significance of the oceans and the role each person plays in conserving the ocean planet for the future. The project is an international network of aquariums, zoos, museums, and conservation organizations.

Rhode Island PIRG
www.ripirg.org
11 South Angell St., #337
Providence, RI 02906
ph: 401-421-6578
fax: 401-331-5266
info@ripirg.org
Educates citizens to participate in the political process, including fisheries reform and other marine issues.

Rhode Island Saltwater Anglers Association
www.risaa.org
6 Arnold Rd.
Coventry, RI 02816
ph: 401-826-2121
fax: 401-826-3546
info@risaa.org
Educates members concerning fishing techniques, fosters sportsmanship, and supports marine conservation and sound management of fisheries resources.

Save the Bay/Narragansett Baykeeper
www.savebay.org
434 Smith St.
Providence, RI 02906
ph: 401-272-3540 x116
fax: 401-273-7153
jtorgan@savebay.org
Seeks to ensure that Narragansett Bay and its watershed are restored and protected from the harmful effects of human activity.

Sierra Club–Rhode Island Chapter

www.sierraclubri.org
21 Meeting St.
Providence, RI 02903
ph: 401-521-4734
fax: 401-521-4001
brooke@sierraclubri.org
Promotes conservation of the natural environment including
the state's coastline by influencing public policy decisions
—legislative, administrative, legal, and electoral.

Surfrider Foundation–Rhode Island

www.risurfrider.org
PO Box 43
Narragansett, RI 02882
ph: 401-364-9444
info@risurfrider.org
Uses conservation, activism, research, and education to
encourage people to protect and enjoy the world's waves
and beaches. The Surfrider Foundation is a nonprofit environmental organization whose membership consists mainly
of surfers. Its scope has expanded to include protecting
wetlands, bird life, and beaches as well as surf spots. It has
sixty-six chapters worldwide.

SOUTH CAROLINA
Aquatic Realm International Research

www.aquaticrealm.org
2435 E. North St., #318
Greenville, SC 29615
ph: 954-252-3751
fax: 954-252-3751
Offers a comprehensive marine Web site.

Charleston Natural History Society

www.homestead.com/cnhsaudubon
C/O The Charleston Audubon Society
PO Box 504
Charleston, SC 29402
ph: 843-795-6934
parula23@aol.com
Conducts educational programs, field trips, marine and terrestrial conservation projects, sponsored research, and
social activities. The society is a chapter of the Audubon
Society, a national network of community-based nature
centers and chapters, scientific and educational programs,
and advocacy on behalf of areas sustaining important bird
populations.

NOAA Coastal Services Center

www.csc.noaa.gov
2234 South Hobson Ave.
Charleston, SC 29405-2413
ph: 843-740-1200
fax: 843-740-1224
csc@csc.noaa.gov

Serves the nation's state and local coastal resource management programs with training, data, fellowships, funding,
and information. It also sponsors a biannual "Coast" conference that attracts a range of coastal managers, academ-
ics, and activists.

Patriots Point Naval and Maritime Museum

www.state.sc.us/patpt
40 Patriots Point Rd.
Mount Pleasant, SC 29464
ph: 843-884-2727
Seeks to develop and improve the Patriots Point area to
provide a place of education and recreation to foster
among the people pride and patriotism in the nation and
its maritime heritage. This includes a growing understanding of our interdependence with the sea.

Ripley's Aquarium

www.ripleysaquarium.com
1110 Celebrity Circle
Myrtle Beach, SC 29577
ph: 800-734-8888
Seeks to provide a top-quality, world-class marine life facility that will foster environmental education, conservation,
and research, while providing entertainment for visitors.

South Carolina Aquarium

www.scaquarium.org
100 Aquarium Wharf
Charleston, SC 29413-9001
ph: 843-720-1990
education@scaquarium.org
Promotes the conservation of South Carolina's habitats and
biodiversity for the enjoyment of future generations of
South Carolinians.

South Carolina Coastal Conservation League

www.scccl.org
PO Box 1765
Charleston, SC 29401
ph: 843-723-8035
fax: 843-723-8308
scccl@charleston.net
Works with communities, businesses, and other conservation and citizen groups to protect the South Carolina coast.

South Carolina Environmental Law Project

www.scelp.org
PO Box 1380
Pawleys Island, SC 29585
ph: 843-527-0078
fax: 843-527-0540
kthomas@scelp.org
Strives to protect South Carolina's coastal and other natural
resources and environment through legal advocacy.

South Carolina League of Women Voters
www.lwvsc.org
PO Box 8453
Columbia, SC 29202
ph: 843-322-6468
lsuggs@imgsc.com
Works to empower citizens to shape better communities
worldwide by encouraging informed and active participation of citizens in government. Is also active in promoting
coastal zone protection.

South Carolina Wildlife Federation
www.scwf.org
2711 Middleburg Dr., Ste. 104
Columbia, SC 29204
ph: 803-256-0670
fax: 803-256-0690
mail@scwf.org
Facilitates effective habitat conservation and respect for
outdoor traditions for current and future generations
through statewide leadership, education, advocacy, and
partnerships.

**Surfrider Foundation–South Carolina,
Charleston Chapter**
www.surfrider.org/charleston
PO Box 841
Folly Beach, SC 29439
ph: 843-588-0211
fax: 843-588-0211
Uses conservation, activism, research, and education to
encourage people to protect and enjoy the world's waves
and beaches. The Surfrider Foundation is a nonprofit environmental organization whose membership consists mainly
of surfers. Its scope has expanded to include protecting
wetlands, bird life, and beaches as well as surf spots. It has
sixty-six chapters worldwide, including two chapters in
South Carolina.

Surfrider Foundation–South Carolina, Myrtle Beach
www.surfrider.org/myrtlebeach
413 Oxner Ct.
Myrtle Beach, SC 29579
dutch852@yahoo.com

Wildlife Action
www.wildlifeaction.com
PO Box 866
Mullins, SC 29574
ph: 843-464-8473
fax: 843-464-8859
info@wildlifeaction.com
Raises public awareness about the ethical responsibility of
each individual to take care of the environment, wildlife,
and natural resources including coastal and marine
resources.

TEXAS
American Society of Limnology and Oceanography
www.aslo.org
5400 Bosque Blvd., Ste. 680
Waco, TX 76710
ph: 254-399-9635
fax: 254-776-3767
business@aslo.org
Promotes scientific stewardship of aquatic resources for the
public interest.

Audubon Society–Southwest Regional Office
www.audubon.org/chapter/tx
901 S. Mopac, Bldg. II, Ste. 410
Austin, TX 78746
ph: 512-306-0225
taustin@audubon.org
Promotes conservation of native habitat for the protection
of birds, other wildlife, and for the enhancement and
appreciation of the environment. The Audubon Society is a
national network of community-based nature centers and
chapters, scientific and educational programs, and advocacy
on behalf of areas sustaining important bird populations.

Audubon Society–Texas Coastal Islands Sanctuary
www.audubon.org/local/sanctuary/tx-coastal
205 N. Carrizo St.
Corpus Christi, TX 78401
ph: 361-884-2634
taustin@audubon.org
Protects more than 11,000 acres on thirty-three islands
along the Gulf Coast including both natural and dredge
spoil islands. Audubon, which controls access, is the primary
organization providing continuous monitoring and research
on breeding and feeding ecology of wading birds on the
Texas coast, including laughing gulls, brown pelicans,
roseate spoonbills, herons, egrets, ibis, black skimmers, and
oystercatchers.

Bayou Preservation Association
www.bayoupreservation.org
PO Box 131563
Houston, TX 77219-1563
ph: 713-529-6443
fax: 713-529-6481
bpa@hic.net
Works to protect and restore the richness and diversity of
public waterways. BPA facilitates collaborative projects and
public awareness about the region's streams and bayous to
foster watershed management, conservation, and recreation along Houston's defining natural resource.

Calhoun County Resource Watch
PO Box 1001
Seadrift, TX 77983
ph: 361-735-4680
wilsonalamobay@aol.com
Fights polluting industries along the Texas coast to protect
local waters, people, fisheries, and wildlife.

Coastal Bend Audubon Society
www.cobirding.com/cbas
3206 Laguna Shores
Corpus Christi, TX 78418
ph: 361-442-9437
tnicolau@stx.rr.com
Promotes conservation of native habitat for the protection
of birds, other wildlife, and for the enhancement and
appreciation of the environment. The Audubon Society is a
national network of community-based nature centers and
chapters, scientific and educational programs, and advocacy
on behalf of areas sustaining important bird populations.

Coastal Bend Bays Foundation
www.baysfoundation.org
PO Box 23025
Corpus Christi, TX 78403
ph: 361-882-3439
fax: 361-882-5625
cbbf@baysfoundation.org
Promotes conservation of freshwater and coastal natural
resources through communication, advocacy, research, and
education.

Coastal Conservation Association–Texas
www.ccatexas.org/ccatexas/Default.asp
6919 Portwest, Ste. 100
Houston, TX 77024
ph: 713-626-4222
ccatx@ccatexas.org
Promotes conservation of the state's marine resources.

Council for Environmental Education
www.c-e-e.org
5555 Morningside Dr., Ste. 212
Houston, TX 77005
ph: 713-520-1936
fax: 713-520-3008
AngeeAcee@aol.com
Provides environmental education programs and services
that promote stewardship of the environment and further
the capacity of learners to make informed decisions.

Dallas World Aquarium
www.dwazoo.com/default.html
1801 N. Griffin
Dallas, TX 75202
ph: 214-720-2224
fax: 214-954-1744
info@dwazoo.com
Seeks to instill the desire to understand the environment
and the importance of conserving and protecting Earth's
resources.

**Environmental Defense–Texas and
Gulf of Mexico Office**
www.environmentaldefense.org
44 East Ave., Ste. 304
Austin, TX 78701
Works to improve management of the Gulf of Mexico reef
fish and shrimp fisheries, protect endangered sea turtles,
and preserve the health of the region's bay and estuarine
ecosystems.

**Galveston Bay Conservation and
Preservation Association (GBCPA)**
www.gbcpa.net
PO Box 323
Seabrook, TX 77586
ph: 281-326-3343
gbcpa@ev1.net
Works to keep the bay waterways and adjacent land areas
suitable as a recreational and living environment. Founded
in 1974, GBCPA is an organization of private citizens.

Galveston Bay Foundation
www.galvbay.org
17324-A Highway 3
Webster, TX 77598
ph: 281-332-3381
fax: 281-332-3153
gbf@galvbay.org
Safeguards Galveston Bay, one of America's most magnificent natural resources.

Golden Triangle Audubon Society
www.goldentriangleaudubon.org/
PO Box 1292
Nederland, TX 77627
Promotes conservation of native habitat for the protection
of birds, other wildlife, and for the enhancement and
appreciation of the environment. The Audubon Society is a
national network of community-based nature centers and
chapters, scientific and educational programs, and advocacy
on behalf of areas sustaining important bird populations.

Gulf of Mexico Foundation

www.gulfmex.org
PMB 51, 5403 Everhart
Corpus Christi, TX 78411
ph: 361-882-3939
fax: 361-882-1262
info@gulfmex.org
Works to ensure a sustainable quality of life for residents of
and visitors to the Gulf coasts.

Harte Research Institute for Gulf of Mexico Studies

Texas A&M University–Corpus Christi
www.hri.tamucc.edu
6300 Ocean Dr.
Corpus Christi, TX 78412
ph: 361-825-2000
fax: 361-825-2770
Supports and enhances the long-term sustainable use of
the Gulf of Mexico by working with local partners in the
United States, Cuba, and Mexico.

HEART (Help Endangered Animals–Ridley Turtles)

www.ridleyturtles.org
PO Box 681231
Houston, TX 77268-1231
ph: 281-444-6204
fax: 281-444-6204
carole@seaturtles.org
Restores and raises sea turtles for release into the sea.

Lower Laguna Madre Foundation

llmf@granderiver.net
PO Box 153
Port Mansfield, TX 78598
ph: 956-944-2387
fax: 956-944-2278
Offers an annual seminar for scientists to report research
and new findings to the public on this south Texas highly-
saline estuary, an annual beach "trash-off," and a water
monitoring program.

Rio Grande Delta Audubon

www.riograndedeltaaudubon.org/
8801 Boca Chica Blvd.
Brownsville, TX 78521
ph: 956-831-4653
fax: 956-831-0147
riorvpark@aol.com
Promotes conservation of native habitat for the protection
of birds, other wildlife, and for the enhancement and
appreciation of the environment. The Audubon Society is a
national network of community-based nature centers and
chapters, scientific and educational programs, and advocacy
on behalf of areas sustaining important bird populations.

Sierra Club–Texas, Coastal Bend Group

http://texas.sierraclub.org/coastalbend/index.html
PO Box 3512
Corpus Christi, TX 78404
ph: 512-852-7938
fax: 512-852-7938
phsuter@aol.com
Uses lawful means to practice and promote the responsible
use of the earth's ecosystems and resources, and to educate and enlist humanity to protect and restore the quality
of the natural and human environment. The Coastal Bend
Group works to protect Padre Island National Seashore,
protect seaturtle populations in the Gulf, and on a number
of other issues.

Sierra Club–Texas, Houston Regional Group

http://texas.sierraclub.org/houston
PO Box 3021
Houston, TX 77253-3021
ph: 713-895-9309
arlenediehl@cs.com

Sierra Club–Texas, Lone Star Chapter

http://texas.sierraclub.org
PO Box 1931
Austin, TX 78767
ph: 512-477-1729
fax: 512-477-8526
lonestar.chapter@sierraclub.org

Sierra Club–Texas, Lower Rio Grande Valley Group

www.lonestar.sierraclub.org/riogrande/index.html
200 East 11th St.
Weslaco, TX 78596
ph: 956-968-1719
jimcyn@hiline.net

Sierra Club–Texas, Padre Island

PO Box 2189
South Padre Island, TX 78597
ph: 956-761-2859
fax: 956-761-2859
scls@igc.apc.org

Surfrider Foundation–Central Texas

PO Box 684126
Austin, TX 78768
ph: 512-415-6816
Uses conservation, activism, research, and education to
encourage people to protect and enjoy the world's waves
and beaches. The Surfrider Foundation is a nonprofit environmental organization whose membership consists mainly of
surfers. Its scope has expanded to include protecting wetlands, bird life, and beaches as well as surf spots. It has sixty-
six chapters worldwide, including two chapters in Texas.

Surfrider Foundation–Texas Chapter

www.surfrider.org/texas/
PO Box 563
Liberty, TX 77575
ph: 936-336-5428
surfpic@swbell.net

Texas General Land Office

www.glo.state.tx.us/coastal.html
PO Box 12873
Austin, TX 78711
ph: 512-475-0773
fax: 512-463-5233
sam.webb@glo.state.tx.us
Enforces the state's Open Beaches Act and works to preserve the Gulf coast for all to use.

Texas Marine Mammal Stranding Network

www.tmmsn.org
4700 Ave. U, Bldg. 3
Galveston, TX 77551
ph: 409-740-4455
fax: 409-741-4321
for strandings: 800-9MAMMAL
tmmsn@tamug.tamu.edu
Promotes understanding and conserving marine mammals
that strand along the Texas Gulf coast. The network
includes more than 2,000 volunteers.

Texas State Aquarium

www.texasstateaquarium.org
2710 North Shoreline
Corpus Christi, TX 78402
ph: 361-881-1200
fax: 361-881-1257
mermaid@txstateaq.org
Provides educationally enriching, entertaining programming
to a diverse audience while promoting and practicing environmental conservation and wildlife rehabilitation.

U.S. PIRG Field Office–Texas

www.texpirg.org/
700 West Ave.
Austin, TX 78701
ph: 512-479-7287
fax: 512-479-0400
info@texpirg.org
Educates citizens to participate in the political process,
including fisheries reform and other marine issues.

VIRGINIA
American Sportfishing Association

www.asafishing.org/asa/
225 Reinekers Ln., Ste. 420
Alexandria, VA 22314
ph: 703-519-9691
fax: 703-519-1872
info@asafishing.org
Safeguards and promotes the enduring social, economic,
and conservation values of sportfishing in America.

Boat U.S.

www.boatus.com/
880 South Pickett St.
Alexandria, VA 22304
ph: 703-823-9550
fax: 703-461-2847
mail@boatus.com
Provides savings, service, and representation, including
information on conservation issues, to millions of boat
owners.

Citizens for a Better Eastern Shore

www.cbes.org
5248 Willow Oak Rd.
PO Box 882
Eastville, VA 23347-0882
ph: 757-678-7157
fax: 757-678-7216
info@cbes.org
Works to build a sustainable community in North Hampton
and Accomack counties on the eastern shore of Virginia
and to protect adjacent waters.

Citizens for the Preservation of Assateague

www.saveassateague.org
PO Box 1111
Assateague Island, VA 23336
ph: 757-336-6811
cpa@saveassateague.org
Opposes the National Park Service's attempts to encourage
development and private enterprise on Assateague Island
National Seashore. The group formed in December 2002.

Coastal Society

www.thecoastalsociety.org
PO Box 25408
Alexandria, VA 22313
ph: 703-933-1599
fax: 703-933-1596
coastalsociety@aol.com
Addresses emerging coastal issues through dialogue, partnerships, and education.

The Cousteau Society

www.Cousteausociety.org
710 Settlers Landing Rd.
Hampton, VA 23669
ph: 757-722-9300
fax: 757-722-8185
cousteau@cousteausociety.org
Educates people to understand, love, and protect the water systems of the planet.

Earth Force

www.earthforce.org
1908 Mount Vernon Ave., 2nd Fl.
Alexandria, VA 22301
ph: 703-299-9400
fax: 703-299-9485
earthforce@earthforce.org
Engages young people as active citizens to improve the environment and their communities now and in the future.

Future Fisherman Foundation

www.futurefisherman.org
225 Reinekers Ln., Ste. 420
Alexandria, VA 22314
ph: 703-519-9691
fax: 703-519-1872
info@futurefishermen.org
Promotes and educates the public about the sport of fishing.

Mariners Museum

www.mariner.org
100 Museum Dr.
Newport News, VA 23606
ph: 757-596-2222
fax: 757-591-7320
community@mariner.org
Features comprehensive information and exhibits on shipping history.

Monitor National Marine Sanctuary

http://monitor.nos.noaa.gov
100 Museum Dr.
Newport News, VA 23606
ph: 757-599-3122
fax: 757-599-7353
monitor@noaa.gov
Preserves the Civil War ironclad, the U.S.S. *Monitor*. One of thirteen National Marine Sanctuaries dedicated to protecting marine environmental and cultural resources.

National Coalition for Marine Conservation

www.savethefish.org
4 Royal St. SE
Leesburg, VA 20175
ph: 703-777-0037
fax: 703-777-1107
hinmank@mindspring.com
Works to conserve ocean fish and their environment.

National Maritime Center

www.uscg.mil/hq/g-m/nmc/web
4200 Wilson Blvd., Ste. 630
Arlington, VA 22203
ph: 800-664-1080
fax: 800-664-1080
Pursues innovative ways to assist the marine community in gaining and using the environmental protection and enforcement services of the U.S. Coast Guard. The center is a field unit of the Coast Guard.

National Taxpayers Union

www.ntu.org
108 N. Alfred St.
Alexandria, VA 22314
ph: 703-683-5700
fax: 703-683-5700
ntu@ntu.org
Informs the public of fiscal and economic policy issues including federal subsidy programs on the coast that are harmful to the coastal and marine environment.

The Nature Conservancy

http://nature.org
4245 North Fairfax Dr., Ste. 100
Arlington, VA 22203-1606
ph: 703-841-5300
fax: 703-841-5300
comment@tnc.org
Preserves plants, animals, and natural communities through land purchases, establishing conservation easements, and other means. The organization has also established its own marine program.

Navy League of the United States
www.navyleague.org
2300 Wilson Blvd.
Arlington, VA 22201-3308
ph: 703-528-1775
fax: 703-356-5760
execdirector@navyleague.org
Supports the men and women of the sea services and their families.

The Ocean Conservancy–Office of Pollution Prevention and Monitoring
www.oceanconservancy.org
432 N. Great Neck Rd., Ste. 103
Virginia Beach, VA 23454
ph: 757-496-0920
fax: 757-496-3207
nmdmp@oceanconservancyva.org
Manages a host of programs with an emphasis on volunteerism, public outreach, and place-based community pollution prevention. The Ocean Conservancy is a national organization dedicated to protecting ocean ecosystems.

Restore America's Estuaries
www.estuaries.org
3801 North Fairfax Dr., Ste. 53
Arlington, VA 22203
ph: 703-524-0248
fax: 703-524-0287
Protects and restores lands and waters essential to the diversity of coastal life. This is the umbrella organization for a number of groups working on watershed, estuarine, and bay protection and restoration issues.

Riverkeeper–James River Association
www.jamesriverassociation.org/riverkeeper
PO Box 919
Mechanicsville, VA 23111
ph: 804-733-6820; 800-366-9229
keeper@jamesriverassociation.org
Provides a strong voice for the James River and watershed on regional and statewide issues related to water quality, land use, and aquatic resources.

Sierra Club–Virginia
www.virginia.sierraclub.org
6 N. 6th St., #102
Richmond, VA 23219
ph: 804-225-9113
fax: 804-225-9114
Explores, enjoys, and protects the planet, including coastal Virginia.

Sirenian International
www.sirenian.org
200 Stonewall Dr.
Fredericksburg, VA 22401-2110
ph: 540-373-8250
caryn@sirenian.org
Pursues manatee and dugong research, education, and conservation through intercultural collaboration.

Society for Conservation Biology
http://conbio.net
4245 N. Fairfax Dr., Ste. 400
Arlington, VA 22203
ph: 703-276-2384
fax: 703-995-4633
info@conbio.org
Develops scientific means for the protection and restoration of the earth's biodiversity with increasing interest and focus on the ocean's collapsing biodiversity.

Surfrider Foundation–Virginia, Virginia Beach
www.surfrider.org/virginiabeach
PO Box 391
Virginia Beach, VA 23458
ph: 757-491-0640
srfvabch@yahoo.com
Uses conservation, activism, research, and education to encourage people to protect and enjoy the world's waves and beaches. The Surfrider Foundation is a nonprofit environmental organization whose membership consists mainly of surfers. Its scope has expanded to include protecting wetlands, bird life, and beaches as well as surf spots. It has sixty-six chapters worldwide.

Underwater Ordnance Recovery
www.Uwuxo.com
PO Box 14003
Norfolk, VA 23518
ph: 757-451-8545
jamesbarton@uwuxo.com
Develops procedures for nondestructive removal of live ordnance to completely eliminate the threat of human ingestion of organic toxins and heavy metals related to decomposing ordnance (such as bombs or explosive shells). Their removal also eliminates the threat to coral reefs and other ecosystems where live-fire military exercises have taken place.

Virginia Beach City Council/Clean the Bay Day Inc.
C/O Chesapeake Bay Foundation
Hampton Roads Office
142 West York St., Ste. 318
Norfolk, VA 23510
ph: 757-622-1964
Organizes a municipal-supported annual cleanup of the bay and beaches.

Virginia Eastern Shorekeeper
PO Box 961
Eastville, VA 23347
ph: 757-678-6182
shorekeeper@verizon.net
Works to build a sustainable community on the eastern
shore of Virginia and to protect adjacent waters.

Virginia Aquarium and Marine Science Center
www.vmsm.com
717 General Booth Blvd.
Virginia Beach, VA 23451
ph: 757-425-3474
Works to increase public knowledge and appreciation of
Virginia's marine environment and to inspire commitment
to preserve its existence.

VIRGIN ISLANDS
Island Resources Foundation
www.irf.org
6292 Estate Nazareth, #100
St. Thomas, VI, USA 00802
ph: 340-775-6225
fax: 340-779-2022
irf@irf.org
Specializes in providing development and environmental
planning assistance to governments and private nonprofit
environmental organizations of small tropical ocean islands.

The Ocean Conservancy–Virgin Islands
Mongoose Junction, Ste. 9
PO Box 1287 Cruz Bay
St. John, VI 00831
ph: 340-776-4701
fax: 340-776-4701
Focuses on creating a unified and thorough approach to
conserving the islands' marine ecosystems and the important living resources they sustain. The Ocean Conservancy
is a national organization dedicated to protecting ocean
ecosystems that uses science-based advocacy, research, and
public education to inform, inspire, and empower people to
speak and act for the oceans.

Virgin Islands Conservation Society
Arawak Bldg., Ste. 3, Gallows Bay
Christiansted, VI 00820
ph: 340-773-1989
fax: 340-773-1989
Seeks to protect the natural and cultural resources of the
Virgin Islands.

Virgin Islands National Park and Biosphere Reserve
1300 Cruz Bay Creek
St. John, VI 00830
ph: 340-776-6201
fax: 340-775-9592
Covers approximately three-fifths of St. John, and nearly all
of Hassel Island in the Charlotte Amalie harbor on St.
Thomas. This park is renowned throughout the world for its
breathtaking beauty.

WASHINGTON
American Cetacean Society–Puget Sound Chapter
www.acspugetsound.org
PO Box 17136
Seattle, WA 98127
ph: 206-781-4860
info@acspugetsound.org
Works to protect whales, dolphins, porpoises, and their
habitats through research, education, and conservation
actions. ACS is the oldest whale conservation organization
on the planet.

Association of Professional Observers
www.apo-observers.org/Observerweb/Sites/
apo_homepage.htm
PO Box 30167
Seattle, WA 98103
ph: 206-547-4228
APO_obs@hotmail.com
Seeks to provide a comprehensive source of information
for federal marine fisheries observers on issues related to
their career and work environment.

Audubon Society–Washington State Office, Olympia
http://wa.audubon.org/new/audubon
PO Box 462
Olympia, WA 98507
ph: 206-786-8020
fax: 206-786-8020
Seeks to conserve and restore natural ecosystems by focusing on birds, other wildlife, and their habitats for the benefit of humanity and the earth's biological diversity. Audubon
is a national network of community-based nature centers
and chapters, scientific and educational programs, and
advocacy on behalf of areas sustaining important bird
populations.

Audubon Society–Washington State Office, Seattle
http://wa.audubon.org/new/audubon
1411 4th Ave., Ste. 920
Seattle, WA 98101
ph: 206-652-2444
Among its statewide priorities are saving Hood Canal,
Puget Sound, and the Northwest Straits.

Audubon Society–Seattle
www.seattleaudubon.org/home.asp
8050 35th Ave. NE
Seattle, WA 98115
ph: 206-523-4483
fax: 206-523-7779
info@seattleaudubon.org

Black Hills Audubon Society
www.blackhillsaudubon.com
PO Box 2524
Olympia, WA 98501
ph: 360-352-7299
jeanmacg@thurston.com

Cascadia Research Collective
www.cascadiaresearch.org
218½ W. 4th Ave.
Olympia, WA 98501
ph: 360-943-7325
calambokidis@cascadiaresearch.org
Conducts research to manage and protect threatened
marine mammals.

Center for Whale Research
www.whaleresearch.com
PO Box 1577
Friday Harbor, WA 98250
ph: 360-378-5835
fax: 360-378-5835
orcasurv@rockisland.com
Studies the social structure and hunting tactics of area
whales and monitors environmental health.

**Citizens for a Healthy Bay/Commencement
Baykeeper**
www.healthybay.org
917 Pacific Ave., Ste. 100
Tacoma, WA 98402
ph: 253-383-2429
scummings@healthybay.org
Seeks to engage citizens to clean up, restore, and protect
Commencement Bay and its surrounding waters and
habitat.

Columbia Riverkeeper
www.columbiariverkeeper.org
PO Box 1254
Bingen, WA 98605
ph: 509-443-2808
crk@columbiariverkeeper.org
Seeks to restore and protect the water quality of the
Columbia River and all life connected to it, from the headwaters to the Pacific Ocean.

Everett Shorelines Coalition
PO Box 13288
Everett, WA 98206
Works to protect, restore, and enhance the shorelines of
the greater Everett area, including those of the Snohomish
River estuary and Port Gardner Bay, through volunteer
stewardship, advocacy, and education.

Federation of Fly Fishers–Washington State Council
www.washingtoncouncilfff.org
PO Box 921
Gig Harbor, WA 98335
ph: 253-265-6162
fax: 253-265-6162
vjblyoung@comcast.net
Promotes concern for aquatic and marine habitats.

Friends of the San Juans
www.sanjuans.org
PO Box 1344
Friday Harbor, WA 98250
ph: 360-378-2319
fax: 360-378-2324
stephanie@sanjuans.org
Seeks to preserve the quality of life for Washington State
land and water resources.

Industrial Shrimp Action Network
www.shrimpaction.com/
14420 Duryea Ln.
Tacoma, WA 98444
ph: 253-539-5272
fax: 253-539-5054
isanet@shrimpaction.org
Conducts campaigns with and assists nongovernmental
organizations worldwide in addressing the impacts on local
communities, economies, and ecosystems caused by large-
scale shrimp aquaculture.

KoosKooskie Fish
www.wildaboutsalmon.com
PO Box 1305
Walla Walla, WA 99362
ph: 888-244-0900
fax: 509-520-8040
kevin@wildaboutsalmon.com
An Internet-assisted direct seafood marketplace that provides independent family fishers a chance to market in a
sustainable manner.

Lummi Tribe

2616 Kwina Rd.
Bellingham, WA 98226
ph: 360-384-1489
fax: 360-380-1850
Promotes an ongoing commitment to marine conservation.
The Lummi Nation has a long history as a fishing tribe.

Mangrove Action Project

www.earthisland.org/map
PO Box 1854
Port Angeles, WA 98362-0279
ph: 360-452-5866
fax: 360-452-5866
mangroveap@olympus.net
Provides solutions to environmental problems relating to
mangrove habitat and promotes citizens advocacy.

Marine Conservation Biology Institute

www.mcbi.org
15805 NE 47th Ct.
Redmond, WA 98052
ph: 425-883-8914
fax: 425-883-8914
mcbiweb@mcbi.org
Works to protect and restore marine life on the West Coast
and around the United States by encouraging research and
training in marine conservation biology, convening scientists, and doing policy research to frame the marine conservation agenda.

Marine Stewardship Council

www.msc.org
2110 N. Pacific St., Ste. 102
Seattle, WA 98103
ph: 206-691-0188
fax: 206-691-0190
info@msc.org
Recognizes well-managed fisheries and promotes consumer
preferences for seafood with the MSC label.

Muckleshoot Tribe

www.muckleshoot.nsn.us
39015 172nd Ave. SE
Auburn, WA 98092
ph: 206-939-3311
fax: 206-939-5311
hr@muckleshoot.nsn.us
Indian Tribe with a rich history in salmon fishing.

National Wildlife Federation–Western Natural Resource Center

www.nwf.org/northwestern
6 Nickerson St., Ste. 200
Seattle, WA 98109
ph: 206-285-8707
fax: 206-285-8698
Educates and advocates to keep wild the very diverse landscapes in the Pacific Northwest; the wildlife species that
depend on these habitats, including marine habitats; and
the people who care about them.

North Olympic Salmon Coalition

www.nosc.org
PO Box 699
Port Townsend, WA 98368
ph: 360-379-8051
Provides community-based support for salmon recovery.

North Sound Baykeeper

www.re-sources.org
1155 N. State St., Ste. 623
Bellingham, WA 98225
ph: 360-733-8307
waters@re-sources.org
Acts as an advocate and educator for marine water quality
for northern Puget Sound.

Northwest Indian Fisheries Commission

www.nwifc.wa.gov
6730 Martin Way E
Olympia, WA 98516
ph: 360-438-1180
fax: 360-753-8659
contact@nwifc.org
Participates in the management of natural resources in
Western Washington, including conservation and restoration of treaty-protected salmon fisheries.

Northwest Straits Commission

www.nwstraits.org
10441 Bayview-Edison Rd.
Mt. Vernon, WA 98273
ph: 360-428-1084
fax: 360-428-1491
info@nwstraits.org
Serves as a sort of "board of directors" for the Northwest
Straits Marine Conservation Initiative. Commission members
represent tribes, the Puget Sound Water Quality Action
Team, and additional appointments by the governor.

Ocean Advocates
3004 N.W. 93rd St.
Seattle, WA 98117
ph: 206-783-6676
fax: 206-783-1799
Felleman@comcast.net
Promotes sound coastal and ocean policies.

The Ocean Conservancy–Washington State Office
www.oceanconservancy.org
2479 Soundview Dr. NE
Bainbridge Island, WA 98110
ph: 206-780-2593
fax: 206-780-2593
Uses science-based advocacy, research, and public education to inform, inspire, and empower people to speak and
act for the oceans. The Ocean Conservancy is a national
organization dedicated to protecting ocean ecosystems. The
organization has a national office in Washington, D.C., and
ten regional offices, including one in Washington State.
Among its major programs are coral reef protection, debris
monitoring, stormwater runoff and pollution, and coastal
cleanup.

Ocean People Resources
PO Box 1692
Kingston, WA 98346
ph: 360-297-6861
fax: 360-297-6861
oceanpeople@peaceaction.org
Seeks to educate others on the need for effective marine
protection and stewardship.

Olympic Coast National Marine Sanctuary
www.ocnms.nos.noaa.gov
115 Railroad Ave. E, Ste. 301
Port Angeles, WA 98362
ph: 360-457-6622
fax: 360-457-6622
Protects a specific marine area of national significance
along Washington's Olympic Coast.

1000 Friends of Washington
www.futurewise.org
1617 Boylston Ave., Ste. 200
Seattle, WA 98122
ph: 206-343-0681
fax: 206-709-8218
Works to keep overdevelopment from consuming farms,
forests, and coastal and rural areas. The organization also
seeks stronger protections for shorelines throughout
the state.

People for Puget Sound
www.pugetsound.org
911 Western Ave.
Seattle, WA 98104
ph: 206-382-7007
fax: 206-382-7007
people@pugetsound.org
Works to protect and restore the health of Puget Sound
and the Northwest Straits through education and action.
The vision of this citizens group is a clean and healthy
sound, teeming with fish and wildlife, cared for by the people who live in the area.

Point Defiance Zoo and Aquarium
www.pdza.org
5400 North Pearl St.
Tacoma, WA 98407
ph: 206-591-5337
fax: 206-591-5337
carolync@tacomaparks.com
Promotes responsible stewardship of the world's natural
resources.

Port Townsend Marine Science Center
www.ptmsc.org/
532 Battery Way
Port Townsend, WA 98368
ph: 800-566-3932; 360-385-5582
fax: 360-385-7248
info@ptmsc.org
Features hands-on exhibits, interpretive programs and
beach walks, shellfish, and seaweed monitoring programs
for visitors to get involved in. The center is located within
Fort Worden State Park on Admiralty Inlet where the Strait
of Juan De Fuca meets Puget Sound.

Project SeaWolf
www.projectseawolf.org
PO Box 929
Marysville, WA 98270
ph: 425-879-4676
projseawolf@earthlink.net
Protects whales and marine wildlife species in the Pacific
Northwest.

Puget Soundkeeper Alliance
www.pugetsoundkeeper.org
5309 Shilshole Ave. NW, Ste. 215
Seattle, WA 98107
ph: 206-297-7002
fax: 206-297-0409
psa@pugetsoundkeeper.org
Protects and preserves Puget Sound by tracking down
toxic pollutants discharged into its waters and identifying
the polluters.

Save Our Wild Salmon

www.wildsalmon.org
424 Third Ave. W, Ste. 100
Seattle, WA 98119
ph: 206-286-4455
fax: 206-286-4455
sos@wildsalmon.org
Returns salmon to their home waters by restoring rivers.

Sea Shepherd Conservation Society

www.Seashepherd.org
PO Box 2616
Friday Harbor, WA 98250
ph: 360-370-5650
fax: 360-370-5651
info@seashepherd.org
Conserves and protects the world's marine wilderness
ecosystems and wildlife species through public education,
investigation, documentation; and, where appropriate, and
where legal authority exists under international law or
under agreement with national governments, enforces violations of the international treaties, laws, and conventions
designed to protect the oceans. Sea Shepherd is known for
its direct action campaigns that have included the sinking
of pirate whalers.

Seattle Aquarium

www.seattleaquarium.org
1483 Alaskan Way
Seattle, WA 98101
ph: 206-386-4300
fax: 206-386-4328
aquarium.programs@seattle.gov
Inspires and encourages interest in Puget Sound and the
Pacific Northwest through its displays, exhibits, and educational programs.

Sierra Club–Washington State, Seattle

http://cascade.sierraclub.org
180 Nickerson, Ste. 103
Seattle, WA 98109
ph: 206-378-0114
fax: 206-378-0114
seawajoyce@attbi.com
Explores, enjoys, and protects the planet, including the natural landscapes and waters of the Pacific Northwest.

Surfrider Foundation–Washington, Northwest Straits

C/O Jen Prince
5860 Milwaukee
Bellingham, WA 98226
nws@surfrider.org
Uses conservation, activism, research, and education to
encourage people to protect and enjoy the world's waves
and beaches. The Surfrider Foundation is a nonprofit environmental organization whose membership consists mainly
of surfers. Its scope has expanded to include protecting
wetlands, bird life, and beaches as well as surf spots. It
has sixty-six chapters worldwide, including three chapters
in Washington.

Surfrider Foundation–Washington, Seattle

www.seattlesurfrider.org/home.html
C/O Sasha Sicks
2172 NW Boulder Way Dr.
Issaquah, WA 98027

Surfrider Foundation–Washington, Westport

http://users.techline.com/nwsc/westportsurfrider/index.htm
PO Box 326
Westport, WA 98595
ph: 360-268-5371
fax: 360-268-5371
westportsurfrider@hotmail.com

The Tulalip Tribes of Washington

www.tulaliptribes-nsn.gov
6700 Totem Beach Rd.
Tulalip, WA 98271
ph: 800-869-8287
Work actively for marine protection including salmon and
fisheries restoration. The Tulalip Tribes, located in the mid-
Puget Sound area of Washington, have a long history as
water people.

University of Washington School of Marine Affairs

www.sma.washington.edu/index.html
3707 Brooklyn Ave. NE
Seattle, WA 98105
ph: 206-543-0106
fax: 206-543-1417
uwsma@u.washington.edu
Seeks to resolve marine problems and conflicts. This is one
of the top ocean policy centers in America.

Washington Environmental Council
www.wecprotects.org
615 Second Ave., Ste. 380
Seattle, WA 98104
ph: 206-622-8103
fax: 206-622-8113
joan@wecprotects.org
Works at the state level to improve and enforce environmental law. Helped win protection of the Everett shorelines
that feed into Puget Sound.

Washington PIRG
www.washpirg.org
3240 Eastlake Ave. E, Ste. 100
Seattle, WA 98102
ph: 206-568-2850
fax: 206-568-2858
info@washpirg.org
Educates citizens to participate in the political process,
including fisheries reform and other marine issues.

Washington Wildlife Federation
www.washingtonwildlife.org
PO Box 1966
Olympia, WA 98507
ph: 360-705-1903
fax: 360-705-1903

Protects wildlife, habitat, and wild places by focusing on
bioregions including the Pacific Northwest Coast and Puget
Trough between the Cascade and Olympic mountains,
which also includes Puget Sound.

Western Washington University
www.ac.wwu.edu/~huxley
516 High St.
Bellingham, WA 98225
ph: 360-650-3520
fax: 360-650-3520
huxley@cc.wwu.edu
Offers a comprehensive program in environmental education, planning, and environmental policy.

Whale Museum
www.whalemuseum.org
PO Box 945
62 First St. N
Friday Harbor, WA 98250
ph: 360-378-4710
fax: 360-378-4710
Promotes stewardship of whales and the Salish Sea
ecosystem.

Government Agencies Dealing with the Oceans

The U.S. government's oversight of the nation's marine activities can make a shark feeding frenzy look like a formal dining affair. For example, if you're concerned about protecting marine mammals, the waters you have to navigate can be highly treacherous.

The National Oceanic and Atmospheric Administration (NOAA) is known for running the weather service, but it is also responsible for federal marine science, sanctuaries, fisheries (beyond state waters), and coastal management (through state agencies).

- The U.S. Fish and Wildlife Service oversees marine mammals such as walruses, manatees, and sea otters.
- NOAA (with input from the federal Marine Mammal Commission) oversees seals, dolphins, and whales. NOAA Fisheries protects sea turtles at sea, while the Fish and Wildlife Service protects them on the beach. If a sea turtle were to get in trouble in the shore break, it could become a jurisdictional dispute.
- The Department of Agriculture oversees the care of captive dolphins and promotes fish farming (as does the U.S. Sea Grant program).

If you're interested in protecting wetlands and estuaries,
- The Army Corps of Engineers protects coastal wetlands (and hands out permits for their destruction) along with the Environmental Protection Agency (EPA).
- The EPA runs the National Estuary Program, which is not to be confused with NOAA's National Estuary and Research Reserves or the Department of Agriculture's Natural Resource Conservation Service, which works to restore wetlands in Louisiana.

If coral reefs are your passion, you need to know that
- NOAA's Marine Sanctuary program protects reefs off the coasts of Georgia, Florida, Texas, Hawaii, and American Samoa.
- The Department of Interior oversees the 200-mile exclusive economic zone (EEZ) where the United States controls all commercial activities and the reefs of U.S. protectorates such as Midway Island.
- The EPA maintains clean water agreements with coral-rich former U.S. protectorates including, Palau and Micronesia.
- The State Department negotiates EEZ rules, coral reef protection, and fishery treaties with other nations.

Is that clear? Of course it isn't, which is why both the Pew Oceans Commission and the U.S. Commission on Ocean Policy say that the environmental threats to America's coasts and oceans cannot be resolved until we have a more unified, effective, ecosystem-based management system for our public seas (based on the precautionary principle of "Do no harm"). Until that happens, if you have marine issues you'd like addressed, your best hope may be to work with seaweed citizen-organizations that already have some experience dealing with government agencies (see Section I). Another approach may be to call the district office of one of your elected officials. Such a call may, at the least, provide you the right name in the right government agency to contact.

You also should remember that government employees are public servants. If you find they are not being helpful, you should insist that they provide you with clear explanations and easy access to the public information you seek.

Of course, this process need not be adversarial. Many dedicated professionals within organizations such as NOAA, the Fish and Wildlife Service, and the EPA choose government service out of a commitment to environmental stewardship and are also frustrated with the way the system is presently organized. Finding these people, and developing relationships with them, can make a difficult journey far more rewarding.

Generally, it's a good idea to begin with your local level of government. If you are concerned about the water quality at a favorite swimming beach, for example, and want to know if health warning signs have been posted, you might call your county department of public health. Though a federal law requires testing of swimming beaches, states administer the law, and local health departments usually do the actual tests and post warning signs. Similarly, if you're concerned about what is running out of your community storm drain or being built on the shores of your favorite cove, it's likely that you're seeing the results of poor planning by your municipal water agency or the influence of short-term interests like coastal developers on your county zoning board. Attending and speaking out at a zoning board hearing is often an effective way to protect yourself and your neighbors from unwise decisions.

Many local decisions (for example, to build or dredge on salt marshes that provide vital natural services, such as cleansing pollutants and providing habitat for wildlife) may also require federal permits from the Army Corps of Engineers and approval from the regional EPA office. Often citizens will find that while one agency (such as the Corps) may be willing to approve a development that threatens a favorite hiking or kayaking spot, another agency (such as Fish and Wildlife) may recognize the ecological value of that spot and become an ally in your fight. For example, a coalition in Savannah, Georgia, that included the city, local business community, environmentalists, and Fish and Wildlife has blocked dredging plans promoted by the local port authority, Army Corps of Engineers, and a local member of Congress.

Presently, few areas exist along our coasts and in our nearshore waters (out to the three-mile state limit) where local, state, tribal, and federal activities don't overlap. If you see someone stealing lobsters from a protected area (or from someone else's traps), you might call your state fish and game warden. If this act occurs in a federal marine sanctuary, either a state or a NOAA enforcement agent might respond. In some areas, such as the Florida Keys National Marine Sanctuary, cross-deputizing Florida Marine Patrol officers to also function as Sanctuary enforcement officers helps eliminate some confusion. And the U.S. Coast Guard stands ready to enforce a range of laws involving everything from oil spills to ship strikes on right whales to guarding our ports against terrorists.

If you have certain overarching concerns, such as protecting the world's coral reefs or whales, you may have to deal with state, federal, United Nations, and other international agencies all at once (although some of those states may not be where you live but where the animals reside). Again, groups that focus on these issues may help you focus your concern and assist you in dealing with the government. Our strength is in our numbers, after all (being right is an added benefit). The following list of federal and state ocean agencies will help you avoid drowning in red tape.

FEDERAL COASTAL AND OCEAN AGENCIES

DEPARTMENT OF AGRICULTURE

The Animal and Plant Health Inspection Service sets standards for zoos and aquariums that keep marine mammals. The Conservation Reserve Enhancement Program provides money to farmers and ranchers who voluntarily plant trees, grasses, and other vegetation along waterways to reduce sediment and nutrient runoff that will eventually find its way into coastal waters. The department also promotes marine aquaculture.

U.S. Dept. of Agriculture
www.usda.gov
1400 Independence Ave. SW
Washington, DC 20250

DEPARTMENT OF COMMERCE

National Oceanic and Atmospheric Administration (NOAA)

NOAA is the main civilian ocean agency on the federal level. Dedicated to predicting and protecting the environment, NOAA's overall mission is twofold: environmental assessment and prediction (observing and assessing the state of the environment while protecting public safety and the nation's economic and environmental security through accurate forecasting); and environmental stewardship (protecting ocean, coastal, and living marine resources while assisting their economic development).

National Oceanic and Atmospheric Administration
www.noaa.gov
Office of Public Affairs
14th St. and Constitution Ave. NW, Room 6217
Washington, DC 20230
ph: 202-482-6090
fax: 202-482-3154
The two main NOAA offices that administer laws, develop regulations, oversee programs, and provide grants for coasts and oceans are the National Ocean Service and the National Marine Fisheries Service.

National Ocean Service (NOS)

The NOS measures and predicts coastal and ocean phenomena, protects large areas of the oceans, works to ensure safe navigation, and serves the American public in many other ways.

National Ocean Service

www.oceanservice.noaa.gov
www.nos.noaa.gov
SSMC4, Room 13632
1305 East-West Highway
Silver Spring, MD 20910
ph: 301-713-3060
fax: 301-713-1213
nos.constituent@noaa.gov
The NOS has eight program offices and two staff offices.

Program offices:
- **Center for Operational Oceanographic Products and Services:**
  http://tidesandcurrents.noaa.gov
- **National Centers for Coastal Ocean Science:**
  http://coastalscience.noaa.gov
- **National Geodetic Survey:**
  http://geodesy.noaa.gov
- **National Marine Sanctuary Program:**
  http://sanctuaries.noaa.gov
- **NOAA Coastal Services Center:**
  http://csc.noaa.gov
- **Office of Coast Survey:**
  http://nauticalcharts.noaa.gov
- **Office of Ocean and Coastal Resource Management:**
  http://coastalmanagement.noaa.gov
- **Office of Response and Restoration:**
  http://response.restoration.noaa.gov

Staff offices:
- **International Program Office:**
  http://international.nos.noaa.gov
- **Management and Budget Office:**
  http://oceanservice.noaa.gov/programs/mb

Office of Ocean and Coastal Resource Management (OCRM)

Within the National Ocean Service, the Office of Ocean and Coastal Resource Management works with a number of coastal states. OCRM is responsible for implementing the Coastal Zone Management Act of 1972, as amended (CZMA). The CZMA created two national programs: the National Estuarine Research Reserve System (see Section Four), and the National Coastal Zone Management Program (CZMP). The CZMP is a unique voluntary federal-state partnership to protect, restore, and responsibly develop the nation's important and diverse coastal communities and resources. Under these programs, states and territories agree to work together to balance conservation and development of coastal resources using state and territorial management authorities, thereby providing for sustainable development of the nation's coasts. The CZMA requires federal permits and other activities to be consistent with enforceable policies adopted by participating states as part of the program. It also authorizes and provides funding for a network of twenty-six national estuarine research reserves.

Office of Ocean and Coastal Resource Management

N/ORM 10th floor SSMC4
1305 East-West Highway
Silver Spring, MD 20910

Coastal States Organization (CSO)

The Coastal States Organization is a Washington, D.C., based nonprofit, nonpartisan association that represents states and territories participating in the CZMP. Formed in 1970, CSO represents the interests of the nation's thirty-five coastal and Great Lakes states and territories to "support the shared vision of the States, Commonwealths, and Territories for the protection, conservation, responsible use, and sustainable economic development of the nation's coastal and ocean resources." Key issues include coastal zone management, offshore and ocean management, coastal hazards, coastal and marine water quality, waterfront revitalization and coastal communities, and public access to the shore. The governors of each state or territory appoint delegates and alternates to represent their interests on the CSO board. Often (but not always) these delegates are the state's program contacts.

Coastal States Organization

www.coastalstates.org
Hall of the States, Ste. 322
444 North Capitol St. NW
Washington, DC 20001
ph: 202-508-3860

STATE COASTAL MANAGEMENT PROGRAM CONTACT INFORMATION

Coastal Zone Management contacts include representatives of different state agencies, departments, and programs who administer the CZMP in their states. While the program is voluntary, the states have a strong incentive to participate both because of the federal grant monies they receive and because of the consistency authority it gives them that requires federal permits and other offshore activities to be consistent with state policies. Nonetheless state Coastal Zone standards vary widely from a high level of coastal environmental protection in California to very low standards in states like Alabama.

ALABAMA
Dept. of Conservation and Natural Resources

Phillip Hinesley
Fairhope, AL 36532
ph: 251-929-0900
fax: 251-990-9293
phinsley@dcnr.state.al.us

ALASKA
Div. of Governmental Coordination

Kerry Howard, Coastal Program Coordinator
PO Box 110030
302 Gold St.
Juneau, AK 99811-0030
ph: 907-465-3562
fax: 907-465-3075
kerry_howard@gov.state.ak.us

AMERICAN SAMOA
Dept. of Commerce

Gene Brighouse-Failauga, Coastal Program Manager
Government of American Samoa
Pago Pago, AS 96799
ph: 684-633-5155 (via Overseas Operator)
fax: 684-633-4195 (via Overseas Operator)
gene.brighouse@czm.noaa.gov

CALIFORNIA

California has three coastal zone management agencies: the San Francisco Bay Conservation and Development Commission (serving the Bay Area), the California Coastal Commission (serving the entire coast), and the California Coastal Conservancy (designated by the state to assure additional coastal protection).

California Coastal Commission

Rebecca Roth, Federal Program Manager
45 Fremont St., Ste. 2000
San Francisco, CA 94105-2219
ph: 415-904-5200
fax: 415-904-5400
rroth@coastal.ca.gov

California Coastal Conservancy

Sam Schuchat, Executive Director
1330 Broadway, 11th Fl.
Oakland, CA 94612-2530
ph: 510-286-4158
fax: 510-286-0470
sschuchat@scc.ca.gov

San Francisco Bay Conservation and Development Commission (BCDC)
William Travis, Executive Director
50 California St., Ste. 2600
San Francisco, CA 94111
ph: 415-352-3600
fax: 415-352-3606
travis@bcdc.ca.gov

CONNECTICUT
Office of Long Island Sound Programs
Charlie H. Evans, Director
Dept. of Environmental Protection
79 Elm St., 3rd Fl.
Hartford, CT 06106-5127
ph: 860-424-3034
fax: 860-424-4054
charles.evans@po.state.ct.us

DELAWARE
Delaware Coastal Management Program
Sarah W. Cooksey, Administrator
Dept. of Natural Resources and Environmental Control
89 Kings Highway
Dover, DE 19901
ph: 302-739-3451
fax: 302-739-2048
scooksey@state.de.us

FLORIDA
Coastal Management Program
Lynn Griffin
Florida Dept. of Environmental Protection
3900 Commonwealth Blvd., Mail Station 47
Tallahassee, FL 32399-3000
ph: 850-992-5438
fax: 850-487-2899

GEORGIA
Coastal Zone Management Program
Stuart Stevens
Georgia Dept. of Natural Resources
One Conservation Way
Brunswick, GA 31520-8687
ph: 912-264-7218
fax: 912-262-3143
stuart@ecology.dnr.state.ga.us

GUAM
Bureau of Planning
Michael Gawel, Administrator
Government of Guam
PO Box 2950
Agana, GU 96910
ph: 671-475-9673
fax: 671-477-1812
mgawel@mail.gov.gu

HAWAII
CZM Program
Chris Chung, Manager
Office of Planning
Dept. of Business, Economic Development, Tourism
PO Box 2359
Honolulu, HI 96804
ph: 808-587-2875
fax: 808-587-2824
cchung@dbedt.hawaii.gov

LOUISIANA
Coastal Management Div.
Terry Howey, Administrator
Dept. of Natural Resources
PO Box 44487
625 North Fourth St.
Baton Rouge, LA 70802
ph: 225-342-7591
fax: 225-342-9439
terryh@dnr.state.la.us

MAINE
State Planning Office/Coastal Program
Kathleen Leyden, Coastal Program Manager
State House Station 38
184 State St.
Augusta, ME 04333-0038
ph: 207-287-3144
fax: 207-287-8059
kathleen.leyden@state.me.us

MARYLAND
Coastal Zone Management Div.
Gwynne Schultz, Director
Dept. of Natural Resources
Tawes State Office Bldg., E-2
580 Taylor Ave.
Annapolis, MD 21401
ph: 410-260-8735
fax: 410-260-8739
gschultz@dnr.state.md.us

MASSACHUSETTS
Executive Office of Environmental Affairs
Tom Skinner, Director
Office of Coastal Zone Management
251 Causeway St., Ste. 900
Boston, MA 02114-2136
ph: 617-626-1201 x400
fax: 617-626-1240
thomas.skinner@state.ma.us

MISSISSIPPI
Office of Coastal Ecology
Daryl Jones, Director
Mississippi Dept. of Marine Resources
1141 Bayview Ave., Ste. 101
Biloxi, MI 39530
ph: 228-374-5000
fax: 228-374-5008
daryl.jones@dmr.state.ms.us

NEW HAMPSHIRE
New Hampshire Coastal Program
David Hartman, Coastal Program Manager
Office of State Planning
2½ Beacon St.
Concord, NH 03301
ph: 603-271-2155
fax: 603-271-1728
dhartman@osp.state.nh.us

NEW JERSEY
New Jersey has two coastal zone management agencies:
one for the entire state coast and one for the salt marsh
meadowlands of northern New Jersey.

New Jersey Dept. of Environmental Protection
Lawrence Schmidt, Director
Office of Coastal Planning and Program Coordination
401 East State St., Box 418
Trenton, NJ 08625-0418
ph: 609-292-2662
fax: 609-292-4608
lschmidt@dep.state.nj.us

**Hackensack–Meadowlands Development
Commission**
Robert Ceberio, Deputy Executive Director
1 DeKorte Park Plaza
Lyndhurst, NJ 07071
ph: 201-460-1700
fax: 201-460-1722
rceberio@hmdc.state.nj.us

NEW YORK
**Div. of Coastal Resources and
Waterfront Revitalization**
George Stafford, Director
Dept. of State
41 State St.
Albany, NY 12231-0001
ph: 518-474-6000
fax: 518-473-2464
gstafford@dos.state.ny.us

NORTH CAROLINA
Dept. of Environment and Natural Resources
Donna Moffitt
Director, Div. of Coastal Management
1638 Mail Service Center
Raleigh, NC 27699-1638
ph: 919-733-2293 x224
fax: 919-733-1495
donna.moffitt@ncmail.net

NORTHERN MARIANAS
Coastal Resources Management
Joaquin Salas, Acting Director
2nd Fl. Morgen Bldg., Box 10007
San Jose Saipan, MP 96950
ph: 670-234-6623
fax: 670-234-0007
crm.director@saipan.com

OREGON
Coastal and Ocean Program Management
Nan Evans
Dept. of Land Conservation and Development
635 Capitol St. NE, Ste. 150
Salem, OR 97301-2540
ph: 503-373-0050
fax 503-731-4068
nan.evans@state.or.us

PENNSYLVANIA
Coastal Zone Management Program
E. James Tabor, Chief
Office for River Basin Cooperation
Dept. of Environmental Protection
PO Box 2063
Harrisburg, PA 17105-2063
ph: 717-772-5626
fax: 717-783-4690
etabor@state.pa.us

PUERTO RICO
Dept. of Natural Resources
Ernesto Diaz, Program Manager
PDA 3½ Ave. Muñoz Rivera
San Juan, PR 00906-6600
ph: 787-721-7593
fax: 787-723-1717
eldiaz@caribe.net

RHODE ISLAND
Coastal Resources Management Council
Grover Fugate, Executive Director
Stedman Office Bldg.
4808 Tower Hill Rd.
Wakefield, RI 02879-1900
ph: 401-783-3370
fax: 401-783-3767
g_fugate@crmc.state.ri.us

SOUTH CAROLINA
Office of Ocean and Coastal Resource Management
Chris Brooks, Deputy Commissioner
South Carolina Dept. of Health and Environmental Control
1362 McMillian Ave., Ste. 400
Charleston, SC 29405-2029
ph: 843-744-5838
fax 803-744-5847
brookscl@chastn86.dhec.state.sc.us

TEXAS
Coastal Div.
Jeb Boyt, Director
Texas General Land Office
1700 North Congress St., #617
Austin, TX 78701
ph: 512-463-5054
fax: 512-475-0680
jeb.boyt@glo.state.tx.us

VIRGINIA
Chesapeake Bay and Coastal Program
Laura McKay
Dept. of Environmental Quality
629 East Main St., 6th Fl.
Richmond, VA 23219
ph: 804-698-4323
fax: 804-698-4319
lbmckay@deq.state.va.us

VIRGIN ISLANDS
Div. of Coastal Zone Management
Janice Hodge, CZM Program Manager
VI Dept. of Natural Resources
Cyrile King Airport, 2nd Fl.
St. Thomas, VI 00802
ph: 340-774-3320
fax: 340-775-5706
janice.hodge@noaa.gov

WASHINGTON
Shorelands and Environmental Assistance Program
Gordon White, Manager
Washington Dept. of Ecology
PO Box 47600
Olympia, WA 98504-7600
ph: 360-407-6966
fax: 360-407-6902
blyn461@ecy.wa.gov

National Marine Fisheries Service (NMFS)
NMFS, recently renamed NOAA Fisheries, is responsible for
the stewardship of living marine resources through science-
based conservation and management and the promotion of
healthy ecosystems (while also promoting the use and
export of U.S. produced seafood).

NOAA Fisheries
www.nmfs.noaa.gov
Director's Office
1315 East-West Highway
Silver Springs, MD 20910-3281
ph: 301-713-2239
Cyber.fish@NOAA.gov

Regional Fishery Management Councils
NMFS works with and oversees eight regional fishery councils that are made up predominantly of fishing industry representatives and who set catch quotas for commercial fish
species in federal waters 3 to 200 miles offshore. The 1976
Magnuson Act, now known as the Magnuson-Stevens Act,
established the councils. As the only federal regulatory bodies exempted from conflict-of-interest laws, the councils'
actions have been highly controversial, with the fishing
industry in effect setting its own catch limits (and with
many commercial species becoming depleted). Both the
Pew Commission and the U.S. Commission on Ocean Policy
have called for reorganization that would have independent conservation scientists set catch quotas and limit the
councils to deciding how that catch would then be divided.

NEW ENGLAND FISHERY MANAGEMENT COUNCIL
The New England Fishery Management Council manages
fisheries in federal waters off the coasts of Maine, New
Hampshire, Massachusetts, Rhode Island, and Connecticut.

New England Fishery Management Council
www.nefmc.org
50 Water St., Mill 2
Newburyport, MA 01950
ph: 978-465-0492
fax: 978-465-3116

MID-ATLANTIC FISHERY MANAGEMENT COUNCIL

The Mid-Atlantic Fishery Management Council manages fisheries in federal waters off the Mid-Atlantic coast of New York, New Jersey, Pennsylvania, Delaware, Maryland, Virginia, and North Carolina (which is represented on both the Mid-Atlantic and South Atlantic councils).

Mid-Atlantic Fishery Management Council
www.mafmc.org
Federal Bldg., Rm. 2115
300 South New St.
Dover, DE 19904-6790
ph: 302-674-2331

SOUTH ATLANTIC FISHERY MANAGEMENT COUNCIL

The South Atlantic Fishery Management Council manages fisheries in federal waters off the coasts of North Carolina, South Carolina, Georgia, and east Florida to Key West.

South Atlantic Fishery Management Council
www.safmc.net
One Southpark Circle, Ste. 306
Charleston, SC 29407-4699
ph: 866-SAFMC-10; 843-571-4366
fax: 843-769-4520

GULF OF MEXICO FISHERY MANAGEMENT COUNCIL

The Gulf of Mexico Fishery Management Council manages fisheries in federal waters off the coasts of Texas to the west coast of Florida.

Gulf of Mexico Fishery Management Council
www.gulfcouncil.org
3018 US Hwy. 301 North, Ste. 1000
Tampa, FL 33619-2266
ph: 813-228-2815

PACIFIC FISHERY MANAGEMENT COUNCIL

The Pacific Fishery Management Council manages fisheries in federal waters off the coasts of Washington, Oregon, and California.

Pacific Fishery Management Council
www.pcouncil.org
2130 SW Fifth Ave., Ste. 224
Portland, OR 97201
ph: 503-326-6352

NORTH PACIFIC FISHERY MANAGEMENT COUNCIL

The North Pacific Fishery Management Council manages fisheries in federal waters off the coast of Alaska.

North Pacific Fishery Management Council
www.fakr.noaa.gov/npfmc
605 West Fourth Ave., Ste. 306
Anchorage, AK 99501-2252
ph: 907-271-2809

WESTERN PACIFIC FISHERY MANAGEMENT COUNCIL

The Western Pacific Fishery Management Council manages fisheries in federal waters off the coasts of U.S. Pacific Islands.

Western Pacific Fishery Management Council
www.wpcouncil.org
1164 Bishop St., Ste. 1400
Honolulu, HI 96813
ph: 808-522-8220

CARIBBEAN FISHERY MANAGEMENT COUNCIL

The Caribbean Fishery Management Council manages fisheries in federal waters (the EEZ) off the coasts of Puerto Rico and the U.S. Virgin Islands.

Caribbean Fishery Management Council
www.caribbeanfmc.com
268 Muñoz Rivera Ave., Ste. 1108
San Juan, PR 00918-2577
ph: 787-766-5926
fax: 787-766-6239

FISHERIES COMMISSIONS

Fisheries commissions predate the federal fishery management councils. They work cooperatively among their members and with the federal government to manage and conserve healthy coastal fishery resources within state waters (out to three miles).

Atlantic States Marine Fisheries Commission
Fifteen Atlantic states from Maine to Florida (including Pennsylvania) formed this body in 1942 in recognition that fish do not adhere to political boundries. It works to coordinate the conservation and management of the states nearshore fishery resources.

Atlantic States Marine Fisheries Commission
www.asmfc.org
1444 Eye St. NW, Sixth Fl.
Washington, DC 20005
ph: 202-289-6400
fax: 202-289-6051

Pacific States Marine Fisheries Commission

Established in 1947, the Pacific States Marine Fisheries Commission works to resolve fishery issues for the Western states of California, Oregon, Idaho, Washington, and Alaska.

Pacific States Marine Fisheries Commission

www.psmfc.org
45 Southeast 82nd, Ste. 200
Gladstone, OR 97027-2522
ph: 503-650-5400
fax: 503-650-5400

Gulf States Marine Fisheries Commission

Established in 1949, the Gulf States Marine Fisheries Commission works to conserve, develop, and make full use of the Gulf of Mexico's fish resources for the states of Texas, Louisiana, Mississippi, Alabama, and Florida.

Gulf States Marine Fisheries Commission

www.gsmfc.org
PO Box 726
Ocean Springs, MS 39566-0726
ph: 228-875-5912
fax: 228-875-6604

MARINE MAMMAL COMMISSION

Title II of the Marine Mammal Protection Act of 1972 created this independent agency of the U.S. government. Its primary focus and duties are to protect and conserve marine mammals.

Marine Mammal Commission

mmc@mmc.gov
4340 East-West Highway, Ste. 905
Bethesda, MD 20814
ph: 301-504-0087
fax: 301-504-0099

DEPARTMENT OF DEFENSE

The U.S. military controls vast tracts of coastal lands, spans the global seas with naval and marine forces, and has huge and often overlooked impacts on America's Blue Frontier.

U.S. Army Corps of Engineers

The Army Corps of Engineers oversees water projects to reduce flooding, dredges ports and harbors, carries out beach replenishment, protects coastal wetlands, and provides permits for filling in wetlands. Under the Estuary Restoration Act of 2000, the Corps has the duty to coordinate estuaries restorations but has received little funding to do so. The Corps is also working on the multibillion-dollar Everglades Restoration project and on a project to restore wetlands in southern Iraq. In a 2004 report, the National Wildlife Federation and Taxpayers for Common Sense identified $12 billion of Corps projects that they claimed wasted taxpayer dollars while threatening 640,000 acres of wetlands and shoreline. The Corps has eight divisions (called Regional Business Centers) and forty-one district offices in the United States, Europe, and Asia.

U.S. Army Corps of Engineers Headquarters

441 G St. NW
Washington, DC 20314–1000
ph: 202-761-0001
fax: 202-761-1683

Division Offices
North Atlantic Division

302 General Lee Ave.
Brooklyn, NY 11252
ph: 718-765-7018
fax: 718 765-7170

South Atlantic Division

U.S. Army Engineer Division, South Atlantic
Rm. 9M15, 60 Forsyth St. SW
Atlanta, GA 30303-8801
ph: 404-562-5011

Lakes and Ohio River Division

PO Box 1159
Cincinnati, OH 45201-1159

Mississippi Valley Division

USACE Mississippi Valley Division
1400 Walnut St.
Vicksburg, MS 39180
ph: 601-634-5757
fax: 601-634-7110

Northwestern Division

220 NW 8th Ave.
Portland, OR 97209
ph: 402-697-2552

Southwestern Division

1100 Commerce St.
Dallas, TX 75242-0216
ph: 214-767-2510

South Pacific Division

U.S. Army Corps of Engineers
333 Market St.
San Francisco, CA 94105-2197
ph: 415-977-8601

Pacific Ocean Division

Bldg. 525
Fort Shafter, HI 96858-5440
ph: 808-438-8319

U.S. Navy

With a budget roughly twelve times the size of the next two largest maritime agencies (NOAA and the Coast Guard) combined, the U.S. Navy plays a major role on America's Blue Frontier and has significant environmental and policy impacts beyond the realm of national defense. Throughout the Cold War the Navy was the major influence on civilian oceanography through its Office of Naval Research that continues today to fund a broad range of marine science. Exempted from many environmental rules during the Cold War, Navy ports and harbors now contain some of the most toxic sediments in America. The Navy has also had periodic conflicts with environmental advocates over dumping of nuclear and chemical wastes, dredging of new port facilities, sighting of certain bombing ranges, and, more recently, over the use of low-frequency active sonar that has been linked to the deaths of deep-ocean whales.

Secretary of the Navy

www.navy.mil
1000 Navy Pentagon
Washington, DC 20350-1000

Naval Historical Center

The official history keeper of the Department of the Navy, the Naval Historical Center is organized into branches according to specialized subject areas. Please direct your inquiry to the appropriate branch.

Naval Historical Center

www.history.navy.mil
901 M St. SE
Washington, DC 20374
ph: 202-433-3224
fax: 202-433-2833
archives@navy.mil

DEPARTMENT OF HEALTH AND HUMAN SERVICES
Food and Drug Administration (FDA)

The FDA issues advisories on seafood such as recent safety advisories on mercury in tuna, shark, and other marine fish. The FDA regulates the drugs and dyes fed to farmed fish. And, as a result of recent legislation, the agency will coordinate with the Centers for Disease Control and the Environmental Protection Agency on links between human health and oceans.

U.S. Food and Drug Administration

www.fda.gov
5600 Fishers Ln.
Rockville, MD 20857-0001
ph: 888-INFO-FDA; 888-463-6332

DEPARTMENT OF HOMELAND SECURITY

Coastal and ocean issues of environment and security are often linked. For example, the same marine sensors used to detect pollution in coastal waters could also be used to track chemical weapons. America's 95,000 miles of coastline represent both a national treasure and a vulnerability. The rapid consolidation of the Department of Homeland Security brought together several agencies that, while vital in times of national crisis, also have a daily role to play on America's Blue Frontier.

Federal Emergency Management Agency (FEMA)

FEMA is located within the Department of Homeland Security. Aside from responding to hurricanes and other coastal disasters, FEMA also provides over $730 billion worth of flood insurance to residents of coastal barrier islands, erosion zones, flood plains, and other at-risk areas. Critics have claimed this insurance has been a driver of risky coastal development since the program was put in place in 1968. In 2004 Congress enacted minor reforms to reduce claims from repetitive risk properties. FEMA has ten regional offices.

Federal Emergency Management Agency

www.fema.gov
500 C St. SW
Washington, DC 20472
ph: 202-566-1600

U.S. Coast Guard

The U.S. Coast Guard, which moved from the Department of Transportation to the Department of Homeland Security in 2003, is a multimission maritime agency. Its tasks include ensuring maritime safety, mobility, security, national defense, and protection of natural resources. The later includes oil spill response and prevention, fisheries enforcement within the 200 -mile U.S. Exclusive Economic Zone (in cooperation with NOAA Fisheries) and foreign vessel inspections. The Coast Guard is organized into nine regional districts (which are not sequentially numbered).

U.S. Coast Guard Headquarters

www.uscg.mil
Public Affairs
2100 Second St. SW
Washington, DC 20593-0001
ph: 202-267- 2100
fax: 202-267-4839

Regional Coast Guard Districts
District 1

408 Atlantic Ave.
Boston, MA 02110
ph: 617-223-8480

District 5
431 Crawford St.
Portsmith, VA 23704
ph: 757-398-6287

District 7
Brickel Plaza Federal Bldg.
909 SE 1st Ave.
Miami, FL 33131
ph: 305-536-5654

District 8
501 Magazine St.
New Orleans, LA 70130
ph: 504-589-6298

District 9
1240 East 9th St.
Cleveland, OH 44199
ph: 216-522-3910

District 11
Coast Guard Island
Alameda, CA 94501
ph: 510-437-3324

District 13
2915 2nd Ave.
Seattle, WA 98174
ph: 206-220-7237

District 14
300 Ala Moana Blvd.
Honolulu, HI 96850
ph: 808-541-2121

District 17
PO Box 3-5000
Juneau, AK 99802
ph: 907-463-2065

DEPARTMENT OF THE INTERIOR

The Department of the Interior is the principal conservation agency charged with protecting and providing access to the nation's natural and cultural heritage, and honoring trust responsibilities to Indian tribes and island communities.

U.S. Fish and Wildlife Service (USFW)

The USFW has multiple programs that impact coastal and saltwater America. The Coastal Program works with other agencies and business, industry, and nonprofit groups to restore marshes and estuaries for migratory waterfowl and shorebirds. As part of that program, the UFWS also administers the Coastal Barrier Resources Act, which excludes certain high-risk coastal sites (like undeveloped barrier islands) from federal development subsidies and incentives. The UFWS also maps and administers these sites, despite pressure from Congress to whittle them down. Its wildlife refuges (see Section Four) protect some of the nation's most pristine coastal environments. The UFWS also oversees the Endangered Species Act, which protects a number of marine species such as manatees and sea turtles.

U.S. Fish and Wildlife Service

www.fws.gov
Dept. of Interior
1849 C St. NW
Washington, DC 20240
ph: 202-208-4717

U.S. Geological Survey (USGS)

The USGS provides scientific information on natural ecosystems, including oceans, to enhance the quality of life of the nation.

U.S. Geological Survey

www.usgs.gov
911 National Center
Reston, VA 20192
ph: 703-648-6600
fax: 703-648-6600

Minerals Management Service

The Minerals Management Service, a bureau in the Department of Interior, is the federal agency that manages the nation's natural gas, oil, and other mineral resources on the Outer Continental Shelf. The agency collects and disburses more than $5 billion per year in revenues from federal offshore mineral leases and from onshore mineral leases on federal and Indian lands.

Minerals Management Service

www.mms.gov
Dept. of Interior
1849 C St. NW
Washington, DC 20240

National Park Service

The National Park Service acts as steward for a number of shoreline national parks, national seashores, and other coastal resources (see Section Four).

National Park Service

Dept. of Interior
1849 C St., NW
Washington, DC 20240

DEPARTMENT OF STATE
Bureau of Oceans and International Environmental and Scientific Affairs (OES)

The OES coordinates an extensive portfolio of issues related to science, the environment, and the world's oceans. The Oceans and Fisheries Directorate includes the Office of Marine Conservation, which focuses on international fisheries and related problems. The Office of Oceans Affairs oversees international ocean law and policy, marine pollution, marine mammals, polar affairs, maritime boundaries, and marine science. One of their initiatives, the "White Water to Blue Water" Partnership, is a Caribbean-wide program to promote integrated watershed and marine ecosystems management even though it doesn't include Cuba.

Bureau of Oceans and International Environmental and Scientific Affairs

www.state.gov/g/oes
U.S. Dept. of State
2201 C St. NW
Washington, DC 20520
ph: 202-647-4000

DEPARTMENT OF TRANSPORTATION
Maritime Administration (MARAD)

MARAD promotes the development and maintenance of the U.S. Merchant Marine, "sufficient to carry the Nation's domestic waterborne commerce and a substantial portion of its waterborne foreign commerce" (though low-paid non-U.S. sailors on foreign-flagged vessels do almost all of this work). MARAD also seeks to ensure that the United States maintains adequate shipbuilding and repair services, efficient ports, effective intermodal water and land transportation systems, and reserve shipping capacity for use in times of national emergency.

Maritime Administration

www.marad.dot.gov
U.S. Dept. of Transportation
400 Seventh St. SW
Washington, DC 20590
ph: 800-99-MARAD; 800-996-2723

ENVIRONMENTAL PROTECTION AGENCY (EPA)

The EPA and the Army Corps of Engineers jointly administer coastal wetlands and tidal marshes under the Clean Water Act (CWA). In addition, the U.S. Fish and Wildlife Service, the National Marine Fisheries Service, and state resource agencies have important advisory roles.

A federal permit is required to discharge dredged or fill material into wetlands and other waters of the United States. The Corps issues the permits for such projects, but the EPA oversees and may veto the permits if the projects put the environment at risk.

Other responsibilities of the EPA include issuing permits for marine and ocean discharges. Under the CWA any industry discharging a pollutant from a point source (for example, a municipal or industrial facility) into U.S. waters must obtain a national pollutant discharge elimination system (NPDES) permit. The EPA also regulates vessel discharges to prevent direct pollution from ships, as well as the introduction of invasive species (from ballast water or by other means).

Ocean dumping cannot occur unless a permit is issued. In the case of (port and harbor) dredged material, the Corps makes the decision to issue a permit, subject to the EPA's concurrence. For all other materials, the EPA is the permitting agency. The agency also designates recommended ocean dumping sites for all types of materials. Under the Clean Water Act, and in cooperation with NOAA under the Coastal Zone Management Act, the EPA leads the (largely unsuccessful to date) effort to eliminate nonpoint sources of ocean pollution, including nitrogen and phosphorous from agriculture, urban runoff, and sewage, which are creating oxygen-deprived dead zones in the Gulf of Mexico and elsewhere around the nation. The EPA oversees the National Estuary Program, the Stormwater Control Program, and the BEACH Act, which helps states test the water quality of recreational beaches.

Environmental Protection Agency

www.epa.gov/owow
Office of Wetlands, Oceans, and Watersheds (4501T)
1200 Pennsylvania Ave. NW
Washington, DC 20460
ph: 202-566-1300

Coastal EPA Offices

These are the offices that administer EPA's coastal water programs, including permitting and inspections within these regions.

EPA Region 1

Serves Connecticut, Maine, Massachusetts, New Hampshire, Rhode Island, Vermont and ten Tribal Nations.

EPA New England Customer Call Center

New England States: 888-372-7341
Outside New England: 617-918-1111

EPA Region 2

Serves New York, New Jersey, Puerto Rico, the U.S. Virgin Islands, and seven Tribal Nations.

Main Regional Office

290 Broadway
New York, NY 10007-1866
ph: 212-637-5000

EPA Region 3
Serves Delaware, the District of Columbia, Maryland, Pennsylvania, Virginia, and West Virginia.

Region 3 Office
1650 Arch St. (3PM52)
Philadelphia, PA 19103-2029
ph: 215-814-5000

EPA Region 4
Serves Alabama, Florida, Georgia, Kentucky, Mississippi, North Carolina, South Carolina, and Tennessee.

Sam Nunn Atlanta Federal Center
61 Forsyth St. SW
Atlanta, GA 30303-3104
ph: 800-241-1754; 404-562-9900
fax: 404-562-8174

EPA Region 6
Serves Louisiana, Arkansas, Oklahoma, New Mexico, Texas, and sixty-six Tribal Nations.

Region 6 Office
1445 Ross Ave., Ste. 1200
Dallas, TX 75202
ph: 214-665-6444

EPA Region 9
Serves Arizona, California, Hawaii, Nevada, the Pacific Islands, and over 140 Tribal Nations.

U.S. EPA Region 9
75 Hawthorne St.
San Francisco, CA, 94105
ph: 866-EPA-WEST; 415-947-8000

EPA Region 10
Serves Alaska, Idaho, Oregon, Washington, and Native Tribes.

U.S. EPA, Region 10
1200 Sixth Ave.
Seattle, WA 98101
ph: 800-424-4EPA; 206-553-1200

NATIONAL AIR AND SPACE ADMINISTRATION (NASA)
NASA funds ocean science research and does remote sensing of oceans from outer space. It also is interested in deep ocean geothermal vent communities, believing that similar life-forms may be discovered in a below-ice ocean on the Jupiter moon, Europa.

NASA Headquarters
www.nasa.gov
Washington, DC 20546-0001
ph: 202-358-0000

NATIONAL SCIENCE FOUNDATION
The National Science Foundation is the major federal sponsor of ocean science research.

National Science Foundation
www.nsf.gov
4201 Wilson Blvd.
Arlington, VA 22230
ph: 703-292-5111

NATIONAL RESEARCH COUNCIL (NRC)
The National Research Council advises the federal government on issues of ocean science, engineering, and policy.

National Research Council
www.nationalacademies.org/osb
2101 Constitution Ave. NW
Washington, DC 20418
ph: 202-334-2714

Ocean Studies Board
The NRC established the Ocean Studies Board to advise the federal government on issues of ocean science, engineering, and policy. In addition to exercising leadership within the ocean community, the board undertakes studies that have often raised the flag about major environmental and other problems confronting the public seas.

Ocean Studies Board
www.nationalacademies.org/osb
2101 Constitution Ave. NW
Washington, DC 20418
ph: 202-334-2714
fax: 202-334-2885

THE WHITE HOUSE
Council on Environmental Quality (CEQ)
The CEQ coordinates federal environmental efforts regarding development of environmental policies including coastal and oceans policies.

Council on Environmental Quality
www.whitehouse.gov/ceq
722 Jackson Pl. NW
Washington, DC 20503
ph: 202-395-5750

STATE MARINE, FISH, AND GAME AGENCIES

Marine, fish, and game agencies are often the lead agencies regulating ocean use and recreational fisheries and enforcing poaching and piracy laws in state waters. Often they work with or provide backup for NOAA Fisheries Enforcement agents, the U.S. Coast Guard, and other federal agencies involved in federal law enforcement (such as protecting migratory birds, marine mammals, or going after wildlife smugglers). Below are the key marine fish and game contacts in the nation's coastal states and territories.

Other important state agencies you might wish to contact include state natural resource agencies not affiliated with the Coastal Zone Management Act or Fish and Game, such as the California Bay-Delta Authority, and state water agencies that are involved in coastal water and watershed activities, such as the South Florida Water Management District.

ALABAMA
Marine Resources Div.
R. Vernon Minton, Director
c/o Alabama Dept. of Conservation and Natural Resources
64 N. Union St.
Montgomery, AL 36130

ALASKA
Alaska Dept. of Fish and Game
Kevin Duffy, Commissioner
PO Box 25526
Juneau, AK 99802-5526

CALIFORNIA
California Dept. of Fish and Game
Ryan Broddrick, Director
PO Box 944209
Sacramento, CA 94244-2090

CONNECTICUT
Bureau of Natural Resources
Edward Parker, Chief
Connecticut Dept. of Environmental Protection
79 Elm St.
Hartford, CT 06106-5127

DELAWARE
Delaware Div. of Fish and Wildlife
Patrick Emory, Director
89 Kings Highway
Dover, DE 19903

FLORIDA
Florida Fish and Wildlife Conservation Commission
Kenneth Haddad, Executive Director
620 S. Meridian St.
Tallahassee, FL 32399-1600

GEORGIA
Wildlife Resources Div.
Noel Holcomb, Director
Georgia Dept. of Natural Resources
2070 U.S. Highway 278 SE
Social Circle, GA 30025

GUAM
Div. of Aquatic and Wildlife Resources
Gerry Davis, Acting Chief
Dept. of Agriculture
192 Dairy Road
Mangilao, GU 96923

HAWAII
Dept. of Land and Natural Resources
Peter Young, Chairperson
State of Hawaii
PO Box 621
Honolulu, HI 96809

LOUISIANA
Louisiana Dept. of Wildlife and Fisheries
Dwight Landreneau, Secretary
PO Box 98000
Baton Rouge, LA 70898-9000

MAINE
Maine Dept. of Marine Resources
21 State House Sta.
Augusta, ME 04333
ph: 207-624-6550

MARYLAND
Maryland Dept. of Natural Resources
Howard King
Director, Fisheries Service
Tawes State Office Bldg.
Annapolis, MD 21401

MASSACHUSETTS
Div. of Fisheries and Wildlife
Wayne MacCallum, Director
Massachusetts Dept. of Fisheries, Wildlife and
Environmental Law Enforcement
One Rabbit Hill Road
Westborough, MA 01581

MISSISSIPPI
Mississippi Dept. of Wildlife, Fisheries, and Parks
Sam Polles, Executive Director
2906 Bldg.
PO Box 451
Jackson, MS 39205

Mississippi Dept. of Marine Resources
1141 Bayview Ave., Ste. 101
Biloxi, MS 39530
ph: 220-374-5000

NEW HAMPSHIRE
New Hampshire Fish and Game Dept.
Lee Perry, Executive Director
11 Hazen Drive
Concord, NH 03301

NEW JERSEY
New Jersey Div. of Fish and Wildlife
Martin McHugh, Director
PO Box 400
Trenton, NJ 08625

NEW YORK
Div. of Fish, Wildlife and Marine Resources
Gerry Barnhart, Director
New York Dept. of Environmental Conservation
625 Broadway, 5th Fl.
Albany, NY 12233-4750

NORTH CAROLINA
North Carolina Wildlife Resources Commission
Charles Fullwood, Executive Director
512 N. Salisbury St.
Raleigh, NC 27604-1188

NORTHERN MARIANAS
Dept. of Lands and Natural Resources
Joaquin a. Tenorio, Secretary
PO Box 10007
Saipan, MP 96950

OREGON
Oregon Dept. of Fish and Wildlife
Lindsay Ball, Director
3406 Cherry Ave. NE
Salem, OR 97303-4924

PUERTO RICO
Puerto Rico Dept. of Natural Resources
Craig G. Lilyestrom
Director, Marine Resources Div.
PDA 3½ Ave., Muñoz Rivera, Puerta de Tierra Station
PO Box 9066600
San Juan, PR 00906-6600

RHODE ISLAND
Rhode Island Div. of Fish and Wildlife
Michael Lapisky, Acting Chief
Stedman Government Center
4808 Tower Hill Rd.
Wakefield, RI 02879

SOUTH CAROLINA
South Carolina Dept. of Natural Resources
John Frampton, Director
PO Box 167
Columbia, SC 29202

TEXAS
Texas Parks and Wildlife Dept.
Robert L. Cook, Executive Director
4200 Smith School Road
Austin, TX 78744

VIRGINIA
Virginia Marine Resources Commission
2600 Washington Ave., Third Fl.
Newport News, VA 23607
ph: 757-247-2200

VIRGIN ISLANDS
Dept. of Planning and Natural Resources
Barbara Kojis
Director, Div. of Fish and Wildlife
6291 Estate Nazareth 101
St. Thomas, VI 00802

WASHINGTON
Washington Dept. of Fish and Wildlife
Jeff Koenings, Director
600 Capitol Way N
Olympia, WA 98501-1091

TRIBAL NATIONS

Indian tribes have soverign treaty rights and should be approached and dealt with as independent governments in any local, regional, or watershed planning that may involve reservation lands and waters.

LOCAL TRIBAL RESOURCE OFFICERS
National Tribal Environmental Council
www.ntec.org
2501 Rio Grande Blvd. NW, Ste. A
Albuquerque, NM 87104
ph: 505-242-2175
fax: 505-242-2654

RESOURCE AND ENFORCEMENT AGENTS
Native American Fish and Wildlife Society
www.nafws.org
8333 Greenwood Blvd., Ste. 250
Denver, CO 80221
ph: 303-466-1725
fax: 303-466-5414

LOCAL GOVERNMENTS

Counties, cities, and townships are perhaps the most vital governmental links ensuring coastal and ocean protection. Urban planning, zoning, and sanitaiton decisions are made at this level.

Elected town and city councils and county supervisors often set the agenda for the kinds of coastal growth that takes place. Zoning boards and community associations set standards that can determine the impact of development on local waters. Departments of sanitation, public works, and/or sewage districts maintain vital infrastructures, including sewage plants, pipes, and storm drains. Public health departments track the safety of swimming waters and locally caught fish. Regional water boards allocate resource use, while local parks departments establish, maintain, and protect recreational coastal areas and beaches. And school boards determine if marine education will be part of your local curriculum.

These and other local government agencies can critically impact the water quality, wildlife habitat, recreational, and other opportunities that will determine whether our coasts and oceans remain healthy and bountiful. Harbor and bay restorations from Boston to Key West to San Francisco to Hilo, Hawaii, show how wise decisions made at the local level can improve both the oceans and the local economies that depend on them.

MARINE SCHOOLS AND SCIENCE CENTERS

Marine scientific studies go back to the earliest days of the Republic, when Benjamin Franklin used the observations of whalers, wind, and temperature readings to identify and chart the Gulf Stream. Still, the nation's first "oceanographic" labs weren't established until the late nineteenth and early twentieth centuries, beginning with the Marine Biological Lab at Woods Hole in 1888; then the Cold Spring Harbor lab in 1890; Stanford University's Hopkins Seaside lab in Monterey, California, in 1891; and the Scripps marine lab down the coast in San Diego, founded in 1903.

Today growing numbers of marine science centers and university programs across the nation provide opportunities for students and scientists alike to explore, better understand, and perhaps help conserve our living ocean planet. With new resources, technologies, and interest, more has been learned about the world's oceans in the last 25 years than in the previous 2,500 years. At the same time, today's exciting new era of coastal and oceanic discovery is taking place alongside an unprecedented assault on our seas from human activities that undermine their ability to function as living ecosystems.

Until World War II, U.S. marine science was mostly boathouse-based biology and zoology focused on the oceans' living resources—fish, invertebrates, marine mammals, and their interactions in the water and along the shore. That focus was expressed in the title of famed Monterey biologist Ed Rickett's classic 1939 text, Between Pacific Tides. The Japanese attack on Pearl Harbor on December 7, 1941, changed the focus as U.S. society, including marine science, mobilized to win the war. Marine science shifted its attention to physical oceanography— the study of sound, salinity, tides, currents, chemistry, and other factors that could help lead to victory at sea (through the development of sonar, underwater explosives, and reliable tidal charts for amphibious landings, among others). When the war ended, the U.S. Navy, through its Office of Naval Research, became the major funder and patron of the nation's marine sciences and would remain so through much of the Cold War. The Navy helped establish or expand marine stations at Woods Hole, Scripps, Princeton, Columbia, the University of Rhode Island, the University of Miami, Texas A&M, Oregon State, and the University of Washington. With their new deep-water fleets and focus on blue water (open ocean) physical oceanography, these centers expanded our understanding of earth processes, even as many biological and ecological aspects of ocean science were neglected.

Since the end of the Cold War, U.S. marine science has been in a state of flux and expansion, bringing the other 71 percent of the planet's surface (and 97 percent of its living habitat) into the study of global change. Some universities and labs have emerged as new centers of multidisciplinary research combining physical, biological, and genetic research (in places like Monterey Bay and the Gulf of Maine), while others, such as Scripps and Woods Hole, continue to play to their traditional strengths with improved tools and vehicles for blue ocean discovery and exploration.

There are, of course, significant differences between different types of schools, programs, and marine stations. A school that focuses on "fisheries science," involving the technologies, biology, economics, and management of fish species targeted for commercial use, will have a different approach from a program in conservation biology that studies the relationship between marine wildlife, habitat, and anthropogenic (human) impacts. Some schools and programs focus on hard science and cutting edge technologies, while others focus on maritime history, anthropology, policy, and law. Major ocean policy schools (which also offer traditional oceanographic programs) include the University of Rhode Island, the University of Delaware, Duke University, and the University of Washington.

Differences are also emerging between generations of marine scientists and scholars. In the past, marine scientists such as Rachel Carson, who spoke out on public policy or tried to popularize their science, were often derided by their colleagues. Many scientists still claim their peer-reviewed empirical work-product should be the sole focus of their careers and leave it to others to draw policy implications from their findings.

But many scientists who received their degrees within the last decade, supported by a handful of older, highly respected leaders in the field, see that the best available science is now producing some of the worst imaginable scenarios in terms of ocean health. They have begun to communicate their findings directly to the public and the media. These younger scientists see quite clearly the need to lead their lives not only as active researchers but also as active citizens of our blue planet.

A list follows, organized by state, of some of the major marine studies programs and research centers located around the United States, including members of the Sea Grant College Program. Like the Land Grant Colleges of the nineteenth century frontier, these schools, with federal and state support, seek to find new and sustainable uses for the U.S. marine environment. We hope this list will provide a starting point for the reader interested in the scientific and academic centers that are now engaged in trying to better understand our Blue Frontier.

MARINE STUDIES PROGRAMS AND RESEARCH CENTERS

ALABAMA
Alabama Center for Estuarine Studies
www.southalabama.edu/aces/
Robert Shipp
LSCB 25, University of South Alabama
Mobile, AL 36688
ph: 251-460-7136
fax: 251-460-7357
rshipp@jaguar1.usouthal.edu
The primary goals of ACES include using sustained experimentation and observation to understand how the most common human-induced modifications of the coastal zone produce changes in ecosystem structure and function; and applying this understanding to develop prudent management strategies for sustaining the productivity of coastal land and seascapes. ACES is affiliated with the University of South Alabama.

Auburn University Marine Extension and Research Center
www.ag.auburn.edu/dept/faa
Richard Wallace
4170 Commanders Dr.
Mobile, AL 36615
ph: 251-438-5690
fax: 251-438-5670
rwallace@acesag.auburn.edu
The center provides fishery-management and business decision information to ensure that marine resources remain renewable.

Dauphin Island Sea Lab
www.disl.org
101 Bienville Blvd.
Dauphin Island, AL 36528
ph: 251-861-2141
fax: 251-861-4646
The Sea Lab is Alabama's marine education and research center. Located on the eastern tip of a barrier island in the Gulf of Mexico, it is the home site of the Marine Environmental Sciences Consortium.

Department of Marine Sciences
www.southalabama.edu/marinesciences/
Robert Shipp
Life Sciences Bldg., Rm. 25
University of South Alabama
Mobile, AL 36688
ph: 251-460-7136
fax: 251-460-7357
agonzale@jaguar1.usouthal.edu
The University of South Alabama offers MS and PhD
degrees in marine sciences with specializations in biological, chemical, and physical geological disciplines.

ALASKA
Alaska Ocean Observing System
www.aoos.org
Molly McCammon
1007 West 3rd Ave., Ste. 100
Anchorage, AK 99501
ph: 907-278-6773
fax: 907-770-6543
mccammon@aoos.org
AOOS seeks to improve science's ability to rapidly detect
and predict changes in marine ecosystems and living
resources.

Coastal Marine Institute
www.sfos.uaf.edu/cmi
School of Fisheries and Ocean Sciences
University of Alaska
Fairbanks, AK 99775-7220
ph: 907-474-6782
fax: 907-474-7204
rpost@sfos.uaf.edu
Created by a cooperative agreement between the University
of Alaska and the U.S. Department of Interior Minerals
Management Service (Alaska Region), the institute studies
coastal topics associated with the development of natural
gas, oil, and minerals in Alaska's outer continental shelf.

**Prince William Sound Science and
Technology Center**
www.pwssc.gen.ak.us
PO Box 705
Cordova, AK 99574
ph: 907-424-5800
frontdes@pwssc.gen.ak.us
The center promotes the goal of maintaining long-term and
self-regulating biodiversity, productivity, and sustainable use
of renewable resources.

School of Fisheries and Ocean Sciences
www.sfos.uaf.edu
University of Alaska–Fairbanks
245 O'Neill Bldg.
Fairbanks, AK 99775
ph: 907-474-7824
fax: 907-474-7204
info@sfos.uaf.edu
The school is responsible for statewide programs relating
to Alaska's vast marine and freshwater environments and
fisheries.

Seward Marine Center
www.ims.uaf.edu:8000/smc/
PO Box 730
Seward, AK 99664
ph: 907-224-5261
A science facility providing access to salt water labs and
the coastal environment.

CALIFORNIA
Bodega Marine Laboratory
www-bml.ucdavis.edu
PO Box 247
Bodega Bay, CA 94923
ph: 707-875-2211
fax: 707-875-2009
ucdbml@ucdavis.edu
The laboratory sits on the windswept headlands of
California's north coast where a 362-acre coastal reserve
meets a state-protected marine reserve. It provides a multidisciplinary approach to solving complex environmental
problems on the marine and terrestrial side of the tideline
in Northern California.

California Maritime Academy
www.csum.edu
William Eisenhardt
200 Maritime Academy Dr.
Vallejo, CA 94590
ph: 707-654-1336
fax: 707-654-1000
The academy provides each student with a quality college
education combining intellectual learning, applied technology, leadership development, and global awareness.

Center for Marine Biodiversity and Conservation
http://cmbc.ucsd.edu/
Scripps Institution of Oceanography
University of California–San Diego
La Jolla, CA 92093-0202
ph: 858-822-2790
fax: 858-822-1267

The center, established in May 2001, strives to meet the challenges of marine conservation through investigation, education, integration, communication, and application of technically sophisticated, regionally appropriate strategies to prevent and reverse biodiversity collapse.

Coastal and Marine Institute

www.scec.sdsu.edu/CMI
Richard Gersberg
San Diego State University
Bldg. HT 03
San Diego, CA 92182
ph: 619-594-2905
fax: 619-594-6112
rgersber@mail.sdsu.edu
The institute studies processes affecting the coastal and marine environment, educates students and the public, and provides advice on the wise use and management of natural resources.

Coastal Studies Center

www.ccs.ucsd.edu
Scripps Institution of Oceanography
University of California–San Diego
La Jolla, CA 92093-0209
ph: 858-534-4333
fax: 858-534-0300
scrippsnews@ucsd.edu
Located adjacent to the SIO pier, the center engages in worldwide scholarly studies of the coastal environment, the development of data acquisition systems and research instrumentation, and advises on coastal protection and sediment management.

Department of Environmental Science and Policy

www.des.ucdavis.edu
University of California–Davis
Davis, CA 95616
ph: 530-752-3026
fax: 530-752-3350
The department offers a comprehensive program in environmental science and policy, including marine issues for both graduate and undergraduate students.

Department of Oceanography

Humboldt State University
1 Harpst St.
Arcata, CA 95521-8299
ph: 707-826-3540
fax: 707-826-4145
Humboldt State is the only campus in the California State University system to offer an undergraduate major in oceanography.

Earth Systems Science and Policy

http://essp.csumb.edu
Sharon Anderson
100 Campus Center, Bldg. 53
California State University–Monterey Bay
Seaside, CA 93955
ph: 831-582-4110
fax: 831-582-3915
sharon_anderson@csumb.edu
The ESSP program offers an innovative interdisciplinary BS degree program linking natural science, physical science, technology, economics, and policy with a focus on watershed and marine environments.

Ecology, Evolution, and Marine Biology

www.lifesci.ucsb.edu/eemb
University of California–Santa Barbara
Santa Barbara, CA 93106
ph: 805-893-4724
fax: 805-893-3511
eemb-info@lifesci.ucsb.edu
EEMB researchers use their scientific understanding to address some of the world's most pressing environmental issues, including the consequences of global warming and harmful algal blooms.

Hopkins Marine Station

www-marine.stanford.edu
Stanford University
Oceanview Blvd.
Pacific Grove, CA 93950
ph: 831-375-0793
fax: 831-655-6200
information@marine.stanford.edu
Hopkins Marine Station is a marine biology research and educational facility that operates as a branch of Stanford University's Department of Biological Sciences. It offers research opportunities and residence courses in a variety of topics related to life in the oceans.

Institute of Marine Sciences

http://ims.ucsc.edu
University of California–Santa Cruz
1156 High St.
Santa Cruz, CA 95064
ph: 831-459-4026
fax: 831-459-2883
IMS provides facilities and administrative and technical support for faculty, researchers, and students interested in marine science and supports a variety of research activities. The institute's on-campus complex includes the IMS office; faculty research laboratories; analytical labs for marine chemistry, biology, and geology; a computer laboratory; a culture room for invertebrates and algae; portable seagoing analytic labs; and support facilities for the Año Nuevo Island program.

Marine Science Center

www.msc.ucla.edu
University of California–Los Angeles
621 Charles E. Young South
Box 951606
Los Angeles, CA 90095
ph: 310-206-8247
Over the past fifty years the marine environment has become of major interest to a large number of UCLA's faculty and students in various disciplines, but historically most of the faculty involved have worked independently. Creation of the MSC facilitates interaction across departmental boundaries of faculty and students with similar interests in marine science. UCLA now offers one of the broadest interdisciplinary educational programs in marine sciences in the United States.

Marine Science Institute

www.msi.ucsb.edu
University of California–Santa Barbara
Santa Barbara, CA 93106
ph: 805-893-8062
fax: 805-893-3765
MSI, established at UC–Santa Barbara in 1969, focuses on marine, coastal zone, and freshwater research; marine policy studies; and educational outreach in marine science. MSI administers and supports research projects involving faculty, professional researchers, technical staff, graduate students, and undergraduate students from fourteen disciplines.

Monterey Bay Aquarium Research Institute

www.mbari.org
7700 Sandholdt Rd.
Moss Landing, CA 95039
ph: 831-775-1700
fax: 831-775-1620
grisel@mbari.org
MBARI focuses on ocean science and technology development through the peer efforts of scientists and engineers. The institute also develops new cutting edge research tools for the oceanographic community.

Moss Landing Marine Laboratories

www.mlml.calstate.edu
8272 Moss Landing Rd.
Moss Landing, CA 95039
ph: 831-771-4400
fax: 831-632-4403
troberts@mlml.calstate.edu
MLML has an international reputation for excellence in marine science research and education. A consortium of seven California State University campuses operates the laboratories, with undergraduate and graduate students taking courses or pursuing MS degrees. The Monterey Submarine Canyon, the largest such feature on the west coast of North America, begins within a few hundred meters of the Moss Landing harbor and the MLML research fleet.

Naval Postgraduate School

www.nps.edu
University Circle
Monterey, CA 93943
ph: 831-656-2441
fax: 831-656-3238
The school emphasizes study and research programs relevant to the Navy's interests, as well as the interests of other branches of the Department of Defense. Admission is open to officers of the armed forces and civilian employees of the U.S. government.

Orange Coast College

www.occ.cccd.edu/~marine/default.html
2701 Fairview Rd.
Costa Mesa, CA 92628
ph: 714-432-5564
tgarriso@mail.occ.cccd.edu
OCC offers a comprehensive undergraduate marine science program.

PRBO Conservation Science

www.prbo.org
4990 Shoreline Hwy.
Stinson Beach, CA 94970
ph: 415-868-1221
fax: 415-868-1946
PRBO Conservation Science is dedicated to conserving birds, other wildlife, and ecosystems through innovative scientific research and outreach. Founded in 1965 as Point Reyes Bird Observatory, its biologists study birds to protect and enhance biodiversity in marine, terrestrial, and wetland systems in western North America.

Romberg Tiburon Center for Environmental Studies

www.rtc.sfsu.edu
3152 Paradise Dr.
Tiburon, CA 94920
ph: 415-338-6063
fax: 415-435-7120
rtcinfo@sfsu.edu
The center performs basic scientific research and educates and trains the next generation of scientists. RTC scientists pursue research in laboratories at the center, at field sites around the world, and through collaborations with colleagues at other universities and institutions.

Scripps Institution of Oceanography

http://sio.ucsd.edu/
Charles Kennel
8602 La Jolla Shores Dr.
La Jolla, CA 92037
ph: 858-453-0167
fax: 858-534-2826
Scripps Institution of Oceanography is one of the oldest, largest, and most important centers for marine science research, graduate training, and public service in the world.

Seymour Marine Discovery Center

www2.ucsc.edu/seymourcenter
100 Shaffer Rd.
Santa Cruz, CA 95060
ph: 831-459-3800
fax: 831-459-1221
The center is part of the Joseph M. Long Marine Laboratory, a research and education facility of the University of California–Santa Cruz. The lab serves as a base for field research in Monterey Bay and the ocean beyond. Institute of Marine Sciences faculty and researchers conduct marine field studies throughout the world.

Wrigley Institute for Environmental Studies

http://wrigley.usc.edu
University of Southern California
AHF 232
Los Angeles, CA 90089
ph: 213-740-6720
fax: 213-740-6780
weis@wrigley.usc.edu
The USC Wrigley Institute for Environmental Studies brings together all of the marine and environmental sciences at the University of Southern California. The jewel in the crown is the unique marine lab on Catalina Island, the Philip K. Wrigley Marine Science Center. Some programs focus entirely on the marine lab and Catalina Island, some on the mainland campus activities, and many are a mix of the two.

Wrigley Marine Science Center

http://wrigley.usc.edu/msc/index.html
PO Box 5069
Avalon, CA 90704
ph: 310-510-4017
fax: 310-510-1364
maureen@usc.edu
This science center, located on Catalina Island, is the marine laboratory for the USC Wrigley Institute for Environmental Studies. It consists of a 30,000 square-foot laboratory building and a dormitory housing and cafeteria complex for up to 24 researchers and groups of up to 60 students. It has a hyperbaric chamber, an administration building, and large waterfront staging areas complete with dock, pier, helipad, and diving lockers.

CONNECTICUT
School of Forestry and Environmental Studies

www.yale.edu/ccws
Yale University
New Haven, CT 06511
ph: 203-432-3026
martha.smith@yale.edu
This leading undergraduate school of the environment also contains a coastal science component that promotes interdisciplinary science and policy studies of watersheds and adjacent coastal waters.

University of Connecticut–Avery Point Campus

www.maritime.uconn.edu
1084 Shennecossett Rd.
Groton, CT 06340
ph: 860-405-9026
fax: 860-405-9075
maritimestudies@uconn.edu
Students at this campus explore the social and cultural side of the human/water relationship through economics, history, literature, and political science.

U.S. Coast Guard Academy

www.cga.edu
R.C. Olsen
31 Mohegan Ave.
New London, CT 06320
ph: 860-444-8444; 860-444-8270
The academy offers education, training, and development for leaders of character who are ethically, intellectually, professionally, and physically prepared to serve the United States and humanity and who also resolve to build on the long military and maritime heritage and proud accomplishments of the U.S. Coast Guard.

DELAWARE
Gerard J. Mangone Center for Marine Policy

www.ocean.udel.edu
University of Delaware
301 Robinson Hall
Newark, DE 19716
ph: 302-831-8086
fax: 302-831-3668
bcs@udel.edu
Marine policy researchers at the University of Delaware work in a number of cross-institutional and cross-national collaborative efforts to develop and disseminate new approaches to the management of ocean and coastal areas. The center is the first of its kind established at an American university.

Graduate College of Marine Studies

www.ocean.udel.edu
University of Delaware
111 Robinson Hall
Newark, DE 19716
ph: 302-831-4389
fax: 302-831-2841
dean@cms.udel.edu
The college aims to advance the knowledge, use, and conservation of global, estuarine, and coastal ocean environments through research, teaching, and service.

DISTRICT OF COLUMBIA
Consortium for Oceanographic Research and Education

www.coreocean.org
Adm. Richard West (Ret.)
1201 New York Ave. NW, Ste. 420
Washington, DC 20005
ph: 202-332-0063
fax: 202-332-8887
core@coreocean.org
CORE is a nonprofit organization based in the District of Columbia that represents sixty-six of the nation's academic institutions, aquariums, nonprofit research institutes, and federal research laboratories.

FLORIDA
Aquaculture Research Center

www.nova.edu/ocean/aqua/index.html
John Reguzzoni
NSU Oceanographic Center
8000 N. Ocean Dr.
Dania Beach, FL 33004
ph: 954-797-1185
fax: 954-262-4098
aquaculture@nova.edu
The center offers research and education in the field of aquaculture (marine fish farming).

Archie Carr Center for Sea Turtle Research

http://accstr.ufl.edu
PO Box 118525
Gainesville, FL 32611
ph: 352-392-5194
fax: 352-392-9166
ACCSTR focuses on sea turtle conservation, research, and education.

Perry Institute for Marine Science

100 North U.S. Highway 1, Ste. 202
Jupiter, FL 33477
ph: 561-741-0192
fax: 561-741-0193
cmrc@cmrc.org

The institute seeks to improve and enhance the understanding of the wider Caribbean region's marine environment by supporting and conducting marine research, education, and conservation programs. Based in Florida since 1970, it also operates research and diving facilities in the Bahamas.

College of Marine Sciences

www.marine.usf.edu
University of South Florida
Doug Myhre
140 7th Ave. South
St. Petersburg, FL 33701
ph: 727-553-1189
fax: 727-553-1130
doug@marine.usf.edu
The college offers a comprehensive graduate research program focused on scientific problem solving in oceanographic environments.

Department of Marine and Environmental Systems

<org-ci>www.fit.edu/AcadRes/dmes/index.html
George Maul
Florida Institute of Technology
150 W. University Blvd.
Melbourne, FL 32901
ph: 321-674-8096
fax: 321-674-7212
dmes@marine.fit.edu
The department seeks to integrate oceanography, ocean engineering, environmental science, meteorology, and related academic concentrations into interdisciplinary knowledge-based optimal solutions to vital contemporary issues through education, research, and service.

Department of Ocean Engineering

www.oe.fau.edu
Florida Atlantic University
777 Glades Rd.
Boca Raton, FL 33431
ph: 561-297-3885
fax: 561-297-3430
sfish@oe.fau.edu
The department provides unique programs in engineering education, research, and technology.

Florida Center for Ocean Sciences Education Excellence

www.fcosee.org
Barbara Spector
140 7th Ave. South
St. Petersburg, FL 33701
ph: 727-553-1189
fax: 813-971-1856
spector@tempest.coedu.usf.edu

The center serves as a regional hub that integrates ocean research with education and outreach activities in Florida and the southeastern United States. It seeks to advance the availability and quality of information on the oceans; improve ocean science competency in an elementary school audience; and keep the public, policy makers, and the media fully informed on ocean issues and discoveries.

Florida Institute of Oceanography

www.marine.usf.edu/FIO/
John Ogden
830 First Street South
St. Petersburg, FL 33701
ph: 727-553-1100
fax: 727-553-1109
FIO works with an array of universities and agencies to support the state's ocean science education and research programs. Using a consortium approach, the institute helps to minimize the expensive duplication of seagoing facilities and equipment needed by the state's academic and agency research and training programs.

Harbor Branch Oceanographic Institution

www.hboi.edu
5600 U.S. 1 North
Fort Pierce, FL 34946
ph: 561-465-2400
fax: 561-465-3644
petri@hboi.edu
The institution is dedicated to exploring the world's oceans, integrating the science and technology of the sea with the needs of humankind. More than 250 scientists, engineers, mariners, and support personnel are involved in research and education in the marine sciences; biological, chemical, and environmental sciences; marine biomedical sciences; marine mammal conservation; aquaculture; and ocean engineering.

Hubbs SeaWorld Research Institute

www.hswri.org
6295 Sea Harbor Dr.
Orlando, FL 32820
ph: 407-370-1651
fax: 407-370-1659
ddefreese@cfl.rr.com
The institute works to ensure that future generations experience the benefits of a healthy environment by conserving the ecological integrity of the oceans and estuaries as a foundation for marine-based economies, sustainable fisheries, public recreation, transportation, tourism, and quality of life.

International Hurricane Research Center

www.ihrc.fiu.edu
Florida International University
11200 SW 8th St.
University Park, MARC 360
Miami, FL 33199
ph: 305-348-1607
fax: 305-348-1605
hurricane@fiu.edu
IHRC, a multidisciplinary center, focuses on the mitigation of hurricane damage to people, property, and the built-up and natural environments. The center has four research laboratories featuring coasts, social science, economics, and wind research.

Marine Laboratory

www.marinelab.fsu.edu
Florida State University
3618 Highway 98
St. Teresa, FL 32358
ph: 850-644-8436
fax: 850-644-2581
bculp@mailer.fsu.edu
The laboratory is located at Turkey Bayou on the Gulf of Mexico, about forty-five miles southwest of Tallahassee. Its facilities include laboratories, classrooms, a modest library, housing, a fleet of small boats, a 47-foot multipurpose research vessel, as well as diving technology and other equipment for education and research in the marine environment.

Marine Science Program/Lab

www.eckerd.edu/academics/nas/msn
Eckerd College
4200 54th Ave. S
St. Petersburg, FL 33711
ph: 800-456-9009; 727-867-1166
smithnf@eckerd.edu
The college's marine science major provides both an integrative science background and specialized foundation work especially suitable for students planning professional careers in marine fields. Students majoring in any track of the marine science major are expected to know fundamental concepts of biological, geological, chemical, and physical oceanography; research methods employed by oceanographers; and history of oceanographic exploration and research.

Maritime Studies

www.uwf.edu/Maritimestudies
Dr. John Bratton, Director
University of West Florida
11000 University Pkwy.
Pensacola, FL 32514
ph: 850-472-2706
Maritimestudies@uwf.edu

The interdisciplinary program prepares students for a variety of occupations through the integration of diverse maritime themes. It integrates five primary fields of study: anthropology and archaeology, environmental studies, history, government, and biology.

Mote Marine Lab

www.mote.org
1600 Ken Thompson Pkwy.
Sarasota, FL 34236
ph: 941-388-4441
fax: 941-388-4441
MML is an independent, nonprofit research organization dedicated to excellence in marine science and education. Founded in 1955 as a place where people could "learn about the sea," much of the laboratory's efforts are directed toward the Southwest Florida coastal region. Through this research, the laboratory provides the exchange of scientific information and hosts visiting investigators, student interns, seminars, and conferences.

Oceanographic Center

www.nova.edu/ocean/
Richard Dodge
NOVA Southeastern University
8000 North Ocean Dr.
Dania Beach, FL 33004
ph: 954-262-4098
fax: 954-262-3600
imcs@nova.edu
The center carries out innovative basic and applied research and provides high-quality graduate and undergraduate education in a broad range of marine science and related disciplines.

Rosenstiel School

www.rsmas.miami.edu
Otis Brown
University of Miami
4600 Rickenbacker Causeway
Miami, FL 33149
ph: 305-361-4711
fax: 305-361-4000
dean@rsmas.miami.edu
The Rosenstiel School is one of the nation's leading graduate schools of marine and atmospheric science.

Seahorse Key Marine Laboratory

www.zoo.ufl.edu/shkml
PO Box 118525
Gainesville, FL 32611
ph: 352-392-1101
fax: 352-392-3704
hbl@zoo.ufl.edu

This University of Florida field station in Cedar Keys National Wildlife Refuge provides support for visiting students, faculty, and scientists doing marine research.

Smithsonian Marine Station

www.sms.si.edu
701 Seaway Dr.
Fort Pierce, FL 34949
ph: 561-465-6630
fax: 561-465-6630
paul@sms.si.edu
This research center specializes in marine biodiversity and ecosystems of Florida. Research focuses on the Indian River Lagoon and the offshore waters of Florida's east-central coast, with comparative studies throughout coastal Florida.

Vero Beach Marine Laboratory

http://research.fit.edu/vbml
150 W. Unversity Blvd.
Melbourne, FL 32901
ph: 321-674-7587
fax: 321-674-7238
jlin@fit.edu
The VBML, formerly the Indian River Marine Science Research Center, is a field laboratory established in 1981 to support marine science research and education for the academic and research programs of the Florida Institute of Technology.

Whitney Laboratory for Marine Bioscience

www.whitney.ufl.edu
University of Florida
9505 Ocean Shore Blvd.
St. Augustine, FL 32080
ph: 904-461-4000
fax: 904-461-4052
paa@whitney.ufl.edu
The lab uses marine organisms to determine how the human body functions and malfunctions.

GEORGIA
Marine Sciences

http://alpha.marsci.uga.edu
William Miller
University of Georgia
Marine Sciences Bldg.
Athens, GA 30602
ph: 706-542-5888
fax: 706-542-4299
bmiller@uga.edu
UGA offers an interdisciplinary department of biological, chemical, and physical oceanography, with special emphasis on coastal and estuarine processes.

Skidway Institute of Oceanography
www.skio.peachnet.edu
Jim Sanders
10 Ocean Science Circle
Savannah, GA 31411
ph: 912-598-2310
fax: 912-598-2400
The institute provides the state of Georgia with a nationally and internationally recognized center of excellence in marine science.

HAWAII
Hawaii Institute of Marine Biology
www.hawaii.edu/HIMB
PO Box 1346
Kaneohe, HI 96744
ph: 808-236-7401
fax: 808-236-7443
The institute, located on twenty-nine-acre Coconut Island in Kaneohe Bay and surrounded by sixty-four acres of coral reef, provides research facilities and offers comprehensive programs covering many disciplines of tropical marine science. Graduate and undergraduate students conducting research at the institute are usually enrolled in the departments of Zoology and Oceanography at the University of Hawaii.

School of Ocean and Earth Science and Technology
www.soest.hawaii.edu
Klaus Keil
University of Hawaii
1680 East West Rd., Post 802
Honolulu, HI 96822
ph: 808-956-9152
Established in 1988, SOEST integrates the University of Hawaii's departments of Oceanography, Geology and Geophysics, Meteorology, and Ocean Engineering; the Hawaii Institute of Geophysics and Planetology; the Hawaii Institute of Marine Biology; the Hawaii Natural Energy Institute; the Sea Grant and Space Grant Programs; the Hawaii Undersea Research Laboratory; the International Pacific Research Center; and the Joint Institute for Marine and Atmospheric Research.

Department of Zoology
www.hawaii.edu/zoology
Sheila Conant
University of Hawaii–Manoa
2538 McCarthy Mall, Edmondson 152
Honolulu, HI 96822
ph: 808-956-8617
fax: 808-956-9812
zoology@hawaii.edu

The department's focus of research for both graduate and undergraduate students is Hawaii's unique natural resources, especially its endemic and indigenous marine and terrestrial animals and their habitats.

Natural Energy Laboratory of Hawaii Authority
http://nelha.org
73-4460 Queen Kaahumanu Highway #101
Kailua-Kona, HI 96740
ph: 808-329-7341
fax: 808-326-3262
nelha@nelha.org
NELHA is a state agency that runs a unique and innovative 870-acre ocean technology park adjacent to the Kona Airport on the Big Island. Its assets include a dual-temperature seawater system pumping water from two- and three-thousand-foot depths for ocean-thermal energy production, cold-water aquaculture, and other uses. It provides resources and facilities for research, education, and commercial activities in an environmentally sound and culturally sensitive manner.

LOUISIANA
Center for River-Ocean Studies
www.tulane.edu/~ceros/present.html
Brent McKee
Tulane University
120 Dinwiddie Hall
New Orleans, LA 70118
ph: 504-862-3167
fax: 504-865-5199
bmckee@tulane.edu
Established in 2003, the center focuses on river-ocean research and education—not only on the Mississippi River–Louisiana Coastal interface but at all scales.

Coastal Ecology Institute
www.lsu.edu/cei
Justic Dubravko
Louisiana State University
1209 Energy Coast and Environment Bldg.
Baton Rouge, LA 70803
ph: 225-578-6515
fax: 225-578-6326
wimberly@lsu.edu
The CEI is a research institute within the School of the Coast and Environment. It provides university-based leadership and scientific expertise in finding solutions to environmental problems affecting coastal and marine environments.

Coastal Environmental Research Laboratory

www.pi.uno.edu/ecosystem.htm
Karen James
University of New Orleans
2000 Lakeshore Dr.
New Orleans, LA 70148
ph: 504-280-4020
fax: 504-280-7396
keramsey@uno.edu
This program focuses on large-scale approaches to coastal areas, using simulation modeling as a research tool for ecosystem analysis. The laboratory functions as a part of the Pontchatrain Institute for Environmental Studies, which also includes the estuary research and coastal plant science laboratories.

Louisiana Universities Marine Consortium

www.lumcon.edu
8124 Hwy. 56
Chauvin, LA 70344
ph: 985-851-2800
fax: 985-851-2874
info@lumcon.edu
LUMCON provides coastal laboratory facilities to Louisiana universities and conducts research and educational programs in the marine sciences.

School of the Coast and Environment

www.cceer.lsu.edu
Russell Chapman
Louisiana State University
E302, Howe-Russell
Baton Rouge, LA 70803
ph: 225-388-6316
The school exists to provide knowledge, technology, and human resources for successful management of natural resources. It is a college-level academic and research unit that also hosts the CREST (Coastal Restoration and Enhancement through Science and Technology) Office, a multi-institutional program dedicated to helping restore coastal habitats in Louisiana and Mississippi.

MAINE

Bigelow Laboratory for Ocean Sciences
www.bigelow.org
Louis Sage
PO Box 475
West Boothbay Harbor, ME 04575-0475
ph: 207-673-9600
fax: 207-633-9641
Bigelow Laboratory focuses on understanding oceans through research and education, including research on the biological productivity of marine food webs.

Darling Marine Center for Research, Teaching, and Service

www.dmc.maine.edu/
193 Clark's Cove Rd.
Walpole, ME 04573
ph: 207-563-3146
fax: 207 563-3119
The center conducts state and federally funded research on the world's oceans, trains graduate students, and offers intensive summer internships to undergraduates.

Gulf of Maine Ocean Observing System

www.gomoos.org
Evan Richert
One Canal Plaza, 7th Fl.
Portland, ME 04112
ph: 207-773-8672
fax: 207-773-0423
GoMOOS offers a state-of-the-art system that provides up-to-date information on weather and oceanographic conditions in the Gulf of Maine.

Maine Maritime Academy

www.mainemaritime.edu
Castine, ME 04420
ph: 207-326-2206
fax: 207-326-2206
admissions@mma.edu
The college, located on the coast of Maine, focuses primarily on marine-related programs, offering studies in engineering, science, management, and marine transportation.

Marine Law Institute

www.mainelaw.maine.edu/mli
University of Maine School of Law
246 Deering Ave.
Portland, ME 04102
ph: 207-780-4355
MLI is the only law school in the Northeast affiliated with a marine policy research program.

School of Marine Sciences

www.marine.maine.edu
University of Maine
341 Aubert Hall
Orono, MA 04469
ph: 207-581-4381
fax: 207-581-4388
davidt@maine.edu
The school offers marine studies from molecular biology and biotechnology to fisheries science, fisheries economics, and anthropology; from marine geology and coastal engineering to aquaculture, marine ecology, and all aspects of oceanography.

MARYLAND
Center for Environmental Science
www.umces.edu
Donald Boesch
University of Maryland
PO Box 775
Cambridge, MD 21613
ph: 410-228-9250
boesch@ca.umces.edu
UMCES maintains a comprehensive program of environmental research, education, and service. Its research spans the globe, with scientists based at the federal partnership program Maryland Sea Grant College or at one of three laboratories across the state: Appalachian Laboratory, Chesapeake Biological Laboratory, and Horn Point Laboratory. Much of its research and science activities focus on the functioning of the Chesapeake Bay and its watershed.

NOAA Office of Science and Technology
www.st.nmfs.gov/st/index.html
William Fox
1315 East-West Highway
Silver Spring, MD 20910
ph: 301-713-2367
william.fox@noaa.gov
The office maintains the integrity of the National Marine Fisheries Service scientific enterprise, includes the service's senior scientist, and coordinates the work of its field research centers.

Smithsonian Environmental Research Center
www.serc.si.edu
PO Box 28
Edgewater, MD 21037
ph: 443-482-2200
fax: 443-482-2380
noblem@si.edu
The center is a global leader in the study of ecosystems and the coastal zone, where land and water meet, and where human populations and their impact are most concentrated.

U.S. Naval Academy
www.usna.edu
121 Blake Rd.
Annapolis, MD 21402-5000
ph: 410-293-1000
The academy offers courses in oceanography, naval architecture, and ocean engineering, as well as seamanship and navigation to future U.S. naval officers.

MASSACHUSETTS
Boston University Marine Program
www.bu.edu/bump
Jelle Atema
7 MBL St.
Woods Hole, MA 02543
ph: 508-289-7950
fax: 508-289-7499
This program offers a hands-on, research-oriented curriculum in marine biology, with emphasis on marine ecology, molecular ecology, behavioral ecology, sensory biology, ichthyology, and oceanography.

Intercampus Graduate School of Marine Sciences and Technology
www.umassmarine.net
c/o Brian Rothschild
University of Massachusetts
706 South Rodney French Blvd.
New Bedford, MA 02744
ph: 508-999-8193
fax: 508-999-8197
brothschild@umassd.edu
The school contributes to the scientific understanding, management, protection, and sustainable economic growth of oceans, the continental shelf, coastal zones, and coastal communities.

Marine Biological Laboratory
www.mbl.edu
7 MBL St.
Woods Hole, MA 02543
ph: 508-548-3705
MBL is an international center for research, education, and training in biology, biomedicine, and ecology.

Marine Policy Center
www.whoi.edu/science/MPC/dept
Woods Hole Oceanographic Institution
Crowell House, MS#41
Woods Hole, MA 02543-1138
ph: 508-289-2449
fax: 508-457-2184
The MPC is the social science research unit of the Woods Hole Oceanographic Institution.

Marine Science Center
www.marinescience.neu.edu
Northeastern University
430 Nahant Rd.
Nahant, MA 01908
ph: 781-581-6076
fax: 781-581-7370

MSC has three principal components: research—focusing on marine life, its diversity and biology, and discovering biotechnological and medical potentials in the sea; education—as a teaching lab for graduate and undergraduate education; and outreach—as a learning center for the general public to enhance awareness and knowledge of the oceans.

Massachusetts Maritime Academy

www.maritime.edu
101 Academy Dr.
Buzzards Bay, MA 02532
ph: 508-830-5000
fax: 508-830-5000
The academy graduates men and women to serve in the maritime industry.

MIT/WHOI Joint Program in Oceanography

http://web.mit.edu/mit-whoi/www
Massachusetts Institute of Technology
77 Massachusetts Ave.
Cambridge, MA 02139
ph: 617-253-9784
fax: 617-253-7544
mit-whoi-www@mit.edu
This graduate program draws on the resources of two major institutions, allowing students access to a range of faculty and facilities, and offers at-sea experience for students seeking an advanced degree in one of five marine disciplines.

MISSISSIPPI

Department of Marine Science

www.usm.edu/marine
University of Southern Mississippi
1020 Balch Blvd.
Stennis Space Center, MS 39529
ph: 800-644-4207; 228-688-3177
fax: 228-688-1121
Marine.Science@usm.edu
The department is home to a multidisciplinary program of graduate study and research in marine environments.

Gulf Coast Research Laboratory

www.usm.edu/gcrl
Harriet Perry
703 East Beach Dr.
PO Box 7000
Ocean Springs, MS 39566
ph: 228-872-4200
fax: 228-872-4204
A public, nonprofit research facility working toward a future of sustainable marine and coastal resources through scientific discovery, development of new technology, and education.

NEW HAMPSHIRE

Center for Coastal and Ocean Mapping/Joint Hydrographic Center

www.ccom-jhc.unh.edu
Abby Archila
University of New Hampshire
24 Colovos Rd.
Durham, NH 03824
ph: 603-862-0839
fax: 603-862-3433
info@ccom.unh.edu
A national center for expertise in ocean mapping and hydrographic sciences.

Cooperative Institute for Coastal and Estuarine Environmental Technology

http://ciceet.unh.edu/index_flash.html
University of New Hampshire
126 Nesmith Hall
Durham, NH 03824
ph: 603-862-0190
fax: 603-862-0190
rlangan@cisunix.unh.edu
CICEET works on the scientific development of innovative technologies that reverse the impact of environmental degradation.

NEW JERSEY

Institute of Marine and Coastal Sciences

www.marine.rutgers.edu
J. Frederick Grassle
Rutgers University
71 Dudley Rd.
New Brunswick, NJ 08901
ph: 732-932-6555 x509
grassle@marine.rutgers.edu
The institute is New Jersey's focal point for education, research, and service efforts in estuarine, coastal, and ocean environments.

New Jersey Marine Sciences Consortium

www.njmsc.org
Sandy Hook Field Station
Bldg. #22
Fort Hancock, NJ 07732
ph: 732-872-1300
NJMSC is an affiliation of twenty-seven colleges, universities, and other organizations whose mission is to serve the state through innovative research, education, and outreach designed to address coastal issues, develop marine technology, promote science-based management policy, and improve science literacy and decision making.

Program in Atmospheric and Oceanic Sciences
www.aos.princeton.edu
GFDL
PO Box 308
Princeton Univeristy
Princeton, NJ 08542
ph: 609-452-6502
This PhD program is affiliated with the Department of
Geosciences and with NOAA's Geophysical Fluid Dynamics
Lab. It takes an interdisciplinary approach to the study of
climate, weather, and the oceans. Their facilities include
supercomputers for the realistic simulation of atmospheric
and oceanic motion.

NEW YORK
Lamont-Doherty Earth Observatory
www.ldeo.columbia.edu
Earth Institute
Columbia University
PO Box 1000, 61 Route 9W
Palisades, NY 10964
ph: 845-359-2931
fax: 845-359-2900
director@ldeo.columbia.edu
LDEO is a leading research institution where more than
200 scientists seek fundamental knowledge about the ori-
gin, evolution, and future of the natural world. It was work
on the mid-Atlantic ridge carried out at LDEO in the 1950s
that confirmed the theory of plate tectonics.

Marine Sciences Research Center
www.msrc.sunysb.edu
David Conover
SUNY–Stony Brook University
Endeaver Hall, Rm. 145
Stony Brook, NY 11794
ph: 631-632-8915
fax: 631-632-8700
dconover@notes.cc.sunysb.edu
The MSRC is the State University of New York's center for
marine and atmospheric research, education, and public
service and the state management office of the Long Island
Sound Study.

Pace Environmental Litigation Clinic
www.law.pace.edu/envclinic
78 N. Broadway, Bldg. E
White Plains, NY 10603
ph: 914-422-4343
fax: 914-422-4343
mpostman@law.pace.edu
The clinic intensively immerses students in environmental
law practice, representing public interest groups including
waterkeepers.

Shoals Marine Lab
www.sml.cornell.edu/
Cornell University
G-14 Stimson Hall
Ithaca, NY 14853-7101
ph: 607-255-3717
fax: 607-255-0742
shoals-lab@cornell.edu
Operated in cooperation with the University of New
Hampshire–Durham, the SML offers a unique opportunity
for students to experience marine science. Located on
Appledore Island in the Gulf of Maine, this near-pristine
environment allows students to study many aspects of
intertidal and subtidal ecology.

U.S. Merchant Marine Academy
www.usmma.edu
300 Steamboat Rd.
Kings Point, NY 11024
ph: 516-773-5888
The academy ensures the availability of trained people for
the nation as shipboard officers and as leaders in the transportation field.

NORTH CAROLINA
Aquarius National Undersea Research Center
www.uncw.edu/aquarius/
Steven Miller
5600 Marvin K. Moss Ln.
Wilmington, NC 28409
ph: 910-962-2300
fax: 910-962-2410
smiller@gate.net
Aquarius is an underwater ocean laboratory located in the
Florida Keys—the only underwater lab presently active in
the world. Research teams working aboard for up to eight
days at a time study a range of marine topics related to the
coral reef environment.

Center for Marine Science
www.uncw.edu/cmsr/
Nancy Stevens
University of North Carolina–Wilmington
5600 Marvin K. Moss Ln.
Wilmington, NC 28409
ph: 910-962-2300
stevensn@uncw.edu
The center provides an environment that fosters a multidisciplinary approach to questions in basic marine research in
fields including oceanography, coastal and wetlands studies, marine biomedical and environmental physiology, and
marine biotechnology and aquaculture.

Coastal Resources Management Program
www.research2.ecu.edu/crm/index.htm
Lauriston King
207 Regsdale Hall
East Carolina University
Greenville, NC 27858
ph: 252-328-2484
kingl@mail.ecu.edu
This PhD program emphasizes an integrated, interdisciplinary approach to coastal studies with a focus on science and public policy.

Marine, Earth, and Atmospheric Sciences Department
Gerald Janowitz
North Carolina State University
PO Box 8208
Raleigh, NC 27695
ph: 919-515-3711
janowitz@ncsu.edu
The department is one of the largest interdisciplinary physical science departments in the United States. It offers graduate and undergraduate degrees.

Marine Sciences Program
www.marine.unc.edu
Marc Alperin
University of North Carolina–Chapel Hill
Dept. of Marine Sciences
12-7 Venable Hall, CB#3300 UNC-CH
Chapel Hill, NC 27599
ph: 919-962-1254
fax: 919-962-1252
alperin@email.unc.edu
The program's diverse and productive coastally oriented faculty and its research and training strengths combine to position it as a leader in providing the necessary scientific expertise to serve the state and the nation in meeting the marine challenges of the twenty-first century. It provides both graduate and undergraduate courses in a range of disciplines, including geological, biological, chemical, and physical oceanography.

Nicholas School of the Environment and Earth Science–Marine Laboratory
www.nicholas.duke.edu/marinelab/
Duke University
135 Duke Marine Lab Rd.
Beaufort, NC 28516-9721
ph: 252-504-7503
fax: 252-504-7648
The Marine Laboratory is a campus of Duke University and a unit within the Nicholas School of the Environment. It provides education and research in basic ocean processes, coastal environmental management, marine biotechnology, and marine biomedicine.

OREGON
College of Oceanic Atmospheric Sciences
www.coas.oregonstate.edu
Mark Abbott
Oregon State University
104 COAS Admin Bldg.
Corvallis, OR 97331
ph: 541-737-2064
fax: 541-737-3504
www@coas.oregonstate.edu
The college seeks to increase knowledge and understanding about oceanic and atmospheric sciences to help Oregonians, the nation, and the world respond to the challenges of a dynamic and changing Earth system.

Hatfield Marine Science Center
http://hmsc.oregonstate.edu/
Oregon State University
2030 SE Marine Science Dr.
Newport, OR 97365
ph: 541-867-0100
fax: 541-867-0138
hmsc@oregonstate.edu
A teaching and research facility located in Newport, Oregon, HMSC plays an integral role in programs of marine and estuarine research and instruction as a lab serving resident scientists, as a base for oceanographic studies, and as a classroom for students.

Oregon Institute of Marine Biology
http://darkwing.uoregon.edu/~oimb/
PO Box 5389
63466 Boat Basin Rd.
Charleston, OR 97420
ph: 541-888-2581
fax: 541-888-3250
mlebow@darkwing.uoregon.edu
The institute has been teaching and conducting research in marine biology on the southern Oregon coast since 1924.

Partnership for Interdisciplinary Studies of Coastal Oceans
www.Piscoweb.org
Oregon State University
Department of Zoology
3029 Cordley Hall
Corvallis, OR 97331
PISCO is a large-scale marine research program that focuses on understanding the nearshore ecosystems of the U.S. West Coast. An interdisciplinary collaboration of scientists from four universities, PISCO integrates long-term monitoring of ecological and oceanographic processes at dozens of coastal sites with experimental work in the lab and field. PISCO's findings are applied to issues of ocean conservation and management and are shared through public outreach and student training programs.

PENNSYLVANIA
Applied Research Laboratory
www.arl.psu.edu
Edward Liszka
Pennsylvania State University
PO Box 30, N. Atherton St.
State College, PA 16804
ph: 814-865-6343
sqk7@psu.edu
The laboratory serves as a university center of research and excellence in naval science and technologies, focusing on undersea missions and related areas.

RHODE ISLAND
Coastal Institute
www.ci.uri.edu/
Peter August
University of Rhode Island
URI Bay Campus Box 36
South Ferry Road
Narragansett, RI 02882
ph: 401-874-6869
fax: 401-874-6513
pete@edc.uri.edu
The institute seeks to advance knowledge and develop solutions to environmental problems in coastal ecosystems.

Graduate School of Oceanography
www.gso.uri.edu
David Farmer
University of Rhode Island
URI Bay Campus Box 52
South Ferry Rd.
Narragansett, RI 02882
ph: 401-874-6889
fax: 401-874-6222
thedean@gso.uri.edu
The GSO offers instruction leading to MA and PhD degrees in biological, chemical, physical, and geological oceanography, as well as in interdisciplinary and related areas such as atmospheric chemistry. GSO now offers a nonthesis degree, the Master of Oceanography (MO).

**Graduate School of Oceanography—
Office of Marine Programs**
www.omp.gso.uri.edu/
Sara Hickox
University of Rhode Island
URI Bay Campus Box 54
South Ferry Rd.
Narragansett, RI 02882
ph: 401-874-6486
fax: 401-874-6211
sara@gso.uri.edu

The Office of Marine Programs focuses efforts in the areas of marine and environmental education and science communications.

Marine Affairs Institute
www.law.rwu.edu
Ralph R. Papitto School of Law
Roger Williams University
Ten Metacom Ave.
Bristol, RI 02809
ph: 401-254-5392
marineaffairs@rwu.edu
The institute conducts research focusing on timely legal and policy issues raised by the development and use of oceans and coastal zones.

SOUTH CAROLINA
Belle W. Baruch Institute for Marine Sciences
www.inlet.geol.sc.edu
Madilyn Fletcher
University of South Carolina
607 EWS Bldg.
Columbia, SC 29208
ph: 803-777-3935
fax: 803-777-5288
The institute offers research on environmental processes in tidal, estuarine, and coastal ocean environments.

Department of Marine Science/Center for Marine and Wetlands Studies
www.coastal.edu/science
Coastal Carolina University
PO Box 261954
Conway, SC 29528
ph: 843-349-2219
Coastal Carolina University offers the largest undergraduate marine science program on the East Coast, along with a graduate program for marine and wetlands studies.

Grice Marine Laboratory
www.cofc.edu/~grice/
College of Charleston
205 Fort Johnson Rd.
Charleston, SC 29412
ph: 843-953-9199
fax: 843-953-9200
The laboratory houses academic programs in marine biology and supports teaching and research in evolutionary biology, marine biogeography, cellular and molecular biology, benthic ecology, immunology, microbial ecology, phytoplankton ecology, environmental physiology, fish systematics, invertebrate zoology, and other marine sciences.

Hollings Marine Laboratory
www.nccos.noaa.gov/about
331 Fort Johnson Rd.
Charleston, SC 29412
ph: 843-762-8737
fax: 843-762-8811
This lab provides science and biotechnology applications to
sustain, protect, and restore coastal ecosystems, emphasizing the linkages between environmental and human health.

**NOAA Center for Coastal Environmental Health and
Biomolecular Research**
www.chbr.noaa.gov
Geoff Scott
219 Fort Johnson Rd.
Charleston, SC 29412
ph: 843-762-8700
fax: 843-762-8500
Geoff.Scott@noaa.gov
CCEHBR provides scientific information needed to manage
and protect coastal resources.

TEXAS
Center for Coastal Studies
www.sci.tamucc.edu/ccs/
Elizabeth Smith
Texas A&M University
Natural Resources Center, Ste. 3200
6300 Ocean Dr.
Corpus Christi, TX 78412
ph: 361-825-2736
fax: 361-825-2770
esmith@falcon.tamucc.edu
The center seeks to increase knowledge and understanding
of the marine ecosystems, habitats, flora, fauna, and
socioeconomics of the Texas coast and Gulf of Mexico
through education and research.

Center for Ports and Waterways
http://tti.tamu.edu/cpw/contacts.stm
Jim Kruse
701 N. Post Oak, Ste. 430
Houston, TX 77024
ph: 713-686-2971
fax: 713-686-5396
j-kruse@ttimail.tamu.edu
CPW focuses on applied research, education, and technology transfer in maritime transportation.

Department of Oceanography
www-ocean.tamu.edu
Robert Stickney
Texas A&M University
College Station, TX 77843
ph: 979-845-6331
fax: 979-845-7211
web@ocean.tamu.edu
The department involves students directly in research
onshore and on ships in all oceans of the world.

Harte Research Institute for Gulf of Mexico Studies
www.hri.tamucc.edu
6300 Ocean Dr.
Corpus Christi, TX 78412
ph: 361-825-2552
fax: 361-825-2552
The institute supports and enhances the long-term sustainable use of the Gulf of Mexico in collaboration with scientists from the United States, Mexico, and Cuba.

**Laboratory for Oceanographic and Environmental
Research**
http://loer.tamug.tamu.edu/
5007 Ave. U
Galveston, TX 77551
ph: 409-740-4476
fax: 409-740-4786
santschi@tamug.tamu.edu
The laboratory is a research arm of the Department of
Marine Science at Texas A&M.

Marine Science Institute
www.utmsi.utexas.edu
750 Channel View Dr.
Port Aransas, TX 78373
ph: 361-749-6711
fax: 361-749-6777
gardner@utmsi.utexas.edu
MSI has active research programs in marine science disciplines, including the physiology, biochemistry, and ecology
of marine plants and animals; dynamics of marine ecosystems; biogeochemistry; mariculture; toxicology; and environmental monitoring. It is the oldest marine research station on the Texas Gulf Coast.

Padre Island Field Research Station
PO Box 181300
Corpus Christi, TX 78740
ph: 361-949-8173
fax: 361-949-8023
The U.S. Geological Survey's Padre Island Field Station conducts a variety of studies, research, and conservation efforts
relating to sea turtles that nest on Texas shores.

Shoreline Environmental Research Facility
www.cbi.tamucc.edu/FacilitiesandEquipment/Serf/default.htm
4305 Waldron Rd.
Corpus Christi, TX 78418
ph: 361-825-2646
fax: 361-825-2715
bonner@cbi.tamucc.edu
SERF is a $5 million facility that simulates coastal environments to study processes controlling contaminate fate, effects, and remediation.

Texas Institute of Oceanography
www.tamug.edu/research/TIO.htm
ocean.tamu.edu/Quarterdeck/QD2.2/TIO/tio.html
PO Box 1675
Galveston, TX 77553
ph: 979-845-7211
fax: 979-845-6331
landrya@tamug.edu
TIO is located at Texas A&M at Galveston and provides the research and technology base for the development of marine-related businesses in the State of Texas and the Gulf of Mexico.

VIRGINIA
Center for Oceans Law and Policy
www.virginia.edu/colp
University of Virginia
580 Massie Rd.
Charlottesville, VA 22903
ph: 434-924-7441
fax: 434-924-7441
The center supports research, education, and discussion of legal and public policy issues relating to the ocean.

Old Dominion University
http://web.odu.edu/webroot/orgs/sci/colsciences
4600 Elkhorn Ave., #406
Norfolk, VA 23529
ph: 757-683-4285
fax: 757-683-4285
celrahha@odu.edu
ODU offers a comprehensive marine science program, including graduate and undergraduate degrees in oceanography and related earth sciences. ODU also operates a research vessel, the *RV Fay Slover*, in the lower Chesapeake Bay.

Virginia Institute of Marine Science
www.vims.edu
L. Donelson Wright
College of William and Mary
PO Box 1346
Gloucester Point, VA 23062
ph: 804-684-7000
fax: 804-684-7097
wright@vims.edu
The institute conducts interdisciplinary research in coastal ocean and estuarine science, educates students and citizens, and provides advisory service to policy makers, industry, and the public.

WASHINGTON
Blakely Island Field Station
www.spu.edu/depts/biology/blakely/
Seattle Pacific University
1 University Dr.
Blakely Island, WA 98222
ph: 360-375-6721; 360-375-6001
fax: 360-375-6224
spu@rockisland.com
The BIFS offers education and research in field-based environmental and physical science and the study of island ecology.

Friday Harbor Laboratories
http://depts.washington.edu/fhl/
University of Washington
620 University Rd.
Friday Harbor, WA 98250
ph: 206-543-1484; 360-378-2165
fax: 206-543-1273
fhladmin@u.washington.edu
Located on San Juan Island, the laboratories research many aspects of marine biology and oceanography.

School of Marine Affairs
www.sma.washington.edu/index.html
University of Washington
3707 Brooklyn Ave. NE
Seattle, WA 98105
ph: 206-543-0106
fax: 206-543-1417
uwsma@u.washington.edu
SMA offers a master's program in marine management and policy.

School of Oceanography
www.ocean.washington.edu/ocean_web/
University of Washington
Box 357940
Seattle, WA 98195
ph: 206-543-5060
The School of Oceanography is a national leader in oceanographic research and education of graduate and undergraduate students. Graduate students work in three areas of
specialization (biological, chemical, and physical oceanography; marine geology; and geophysics) and on a variety of
interdisciplinary topics, including climate change, extreme
environments, and coastal systems.

Shannon Point Marine Center
www.ac.wwu.edu/~spmc/
Western Washington University
1900 Shannon Point Rd.
Anacortes, WA 98221
ph: 360-293-2188; 360-650-7400
fax: 360-293-1083
spmc@cc.wwu.edu
The SPMC supports and promotes the academic mission of
Western Washington University with respect to its specialized programs in the marine and estuarine sciences for
both graduate and undergraduate students.

SEA GRANT CONSORTIUMS

The Sea Grant College Program is composed of thirty universities conducting scientific research, education, and
extension projects to enhance the practical use and conservation of coastal, marine, and Great Lakes resources. The
following list provides information on the coastal and
marine centers.

National Sea Grant Program
www.seagrantnews.org
NOAA Headquarters
1315 East-West Highway
Silver Spring, MD 20910-3226
ph: 301-713-2483
fax: 301-713-2483
jana.goldman@noaa.gov
Sea Grant is a partnership and a bridge between government, academia, industry, scientists, and private citizens to
help Americans understand and sustainably use the Great
Lakes and ocean waters for long-term economic growth.
The national program unites 30 state Sea Grant programs,
more than 200 universities, and millions of people.

ALASKA
Alaska Sea Grant
www.uaf.edu/seagrant/
Brian Allee
University of Alaska–Fairbanks
PO Box 755040
Fairbanks, AK 99775-5040
ph: 907-474-7949
fax: 907-474-6285
allee@sfos.uaf.edu

CALIFORNIA
California Sea Grant
www.csgc.ucsd.edu/
Russell Moll
University of California–San Diego
9500 Gilman Dr.
La Jolla, CA 92093
ph: 858-534-4440
fax: 858-534-2231
rmoll@ucsd.edu

University of Southern California
www.usc.edu/org/seagrant
USC Sea Grant University Park
Los Angeles, CA 90089
ph: 213-740-1961
seagrant@usc.edu

CONNECTICUT
Connecticut Sea Grant
www.seagrant.uconn.edu/
Edward Monahan
University of Connecticut
1080 Shennecossett Rd.
Groton, CT 06340
ph: 860-405-9110
fax: 860-405-9109
sgoadm01@uconnvm.uconn.edu

DELAWARE
Delaware Sea Grant
www.ocean.udel.edu/seagrant/
Carolyn Thoroughgood
Graduate College of Marine Studies
University of Delaware
111 Robinson Hall
Newark, DE 19716
ph: 302-831-2841
fax: 302-831-4389
ctgood@udel.edu

FLORIDA
Florida Sea Grant
www.flseagrant.org/
James Cato
University of Florida
Bldg. 803, McCarty Dr.
PO Box 110400
Gainesville, FL 32611
ph: 352-392-5870
fax: 352-392-5113
jcato@mail.ifas.ufl.edu

GEORGIA
Georgia Sea Grant
www.marsci.uga.edu/gaseagrant/
Mac Rawson
University of Georgia
220 Marine Sciences Bldg.
Athens, GA 30602
ph: 706-542-6009
fax: 706-542-3652
mrawson@uga.cc.uga.edu

HAWAII
Hawaii Sea Grant
www.soest.hawaii.edu/SEAGRANT/
E. Gordon Grau
University of Hawaii
2525 Correa Road, HIG 238
Honolulu, HI 96822
ph: 808-956-7031
fax: 808-956-3014
sg-dir@soest.hawaii.edu

LOUISIANA
Louisiana Sea Grant
www.laseagrant.org/
Jack Van Lopik
Louisiana State University
239 Sea Grant Bldg.
Baton Rouge, LA 70803
ph: 225-578-6710
fax: 225-578-6331
jvl@lsu.edu

MAINE
Maine Sea Grant
www.seagrant.umaine.edu/
Paul Anderson
University of Maine
5715 Coburn Hall, Rm. 14
Orono, ME 04469
ph: 207-581-1422
fax: 207-581-1426
panderson@maine.edu

MARYLAND
Maryland Sea Grant College
www.mdsg.umd.edu
Jonathan Kramer
University of Maryland
4321 Hartwick Rd., Ste. 300
College Park, MD 20740
ph: 301-403-4220
fax: 301-403-4255
kramer@mdsg.umd.edu

MASSACHUSETTS
MIT Sea Grant
http://web.mit.edu/seagrant/
Chryssostomos Chryssostomidis
Massachusetts Institute of Technology
Bldg. E38, Rm. 330, Kendall Square
292 Main St.
Cambridge, MA 02139
ph: 617-253-7131
fax: 617-258-5730
chrys@mit.edu

WHOI Sea Grant Program
www.whoi.edu/seagrant/
Judith McDowell
Woods Hole Oceanographic Institution
193 Oyster Pond Road, MS #2
Woods Hole, MA 02543
ph: 508-289-2557
fax: 508-457-2172
jmcdowell@whoi.edu

VIRGINIA
Virginia Sea Grant
www.virginia.edu/virginia-sea-grant/
William Rickards
University of Virginia
Madison House
170 Rugby Rd.
PO Box 400146
Charlottesville, VA 22904
ph: 434-924-5965
fax: 434-982-3694
rickards@virginia.edu

WASHINGTON
Washington Sea Grant
www.wsg.washington.edu/
Louie Echols
University of Washington
3716 Brooklyn Ave. NE
PO Box 355060
Seattle, WA 98105
ph: 206-543-6600
fax: 206-685-0380
echols@u.washington.edu

U.S. COASTAL AND OCEAN PARKS

The United States is blessed with 95,000 miles of coastline, which includes some of the finest coastal parks, beaches, and marine reserves in the world. These are places where you can go to relax, have fun, or be inspired by the wonders of our blue ocean planet. Some, such as Point Reyes National Seashore in California, are easily accessible for day trips or weekend getaways, while others, such as the Northwest Hawaiian Islands Coral Reef Ecosystem Reserve, are among the most remote locations on Earth. What they share in common is that they belong to all of us. A list of national (federal) marine parks and seashores follows. There are millions of acres more of state and local marine parks and beaches. For information on these salty jewels, contact your state, county, or municipal parks departments.

NATIONAL MARINE SANCTUARIES

The U.S. Congress created America's National Marine Sanctuary Program in 1972, 100 years after dedicating Yellowstone as the first national park. Today, thirteen marine sanctuaries exist, including the historic shipwrecks of Thunderbay in Lake Michigan. The remote northwestern Hawaiian Ecosystem Reserve will become the fourteenth sanctuary in the near future. An extension of the Hawaiian Islands, the reserve stretches over 1,200 miles in length and contains 70 percent of America's coral reefs.

Unfortunately, some marine sanctuaries are more akin to heavily logged U.S. national forests than to national parks. While oil drilling and waste dumping are banned within sanctuary boundaries, other extractive and invasive activities are permitted, including commercial fishing, oil shipping, and cable laying. In the early years, Yellowstone's major attractions included hunting elk and buffalo, before people came to fully value terrestrial wilderness. Our hope is that people are now beginning to fully value marine wilderness.

For additional information on the Marine Sanctuary system go to www.sanctuaries.nos.noaa.gov.

AMERICAN SAMOA
Fagatele Bay
This remote sanctuary lies nestled in an eroded volcanic crater on the southwest shore of the island of Tutuila, American Samoa. An ancient Polynesian culture has served as steward of this fringing coral reef ecosystem rich in reef fish, giant clams, and blacktip reef sharks. In the late 1970s, a crown-of-thorns starfish attack devastated the bay, destroying over 90 percent of the coral. Two hurricanes, tropical storms, and coral bleaching followed this incident, each further stressing the reefs. Despite these damaging natural events, the coral has proved resilient, providing a valuable opportunity for scientists to study how tropical reefs recover from such episodes.

Sanctuary designated: April 29, 1986
Protected area: 0.25 square miles
Key species: Crown-of-thorns starfish, blacktip reef shark, surgeonfish, hawksbill turtle, parrotfish, and giant clam
Key habitats: Tropical coral reef

Sanctuary manager: Nancy Daschbach
Headquarters address:
PO Box 4318
Pago Pago, AS 96799
ph: 684-633-7354
fax: 684-633-7355
fagatelebay@noaa.gov
http://fagatelebay.noaa.gov

CALIFORNIA
The Channel Islands
A fertile combination of warm and cool currents converging around five islands makes up the Channel Islands National Marine Sanctuary. This confluence attracts a great variety of plants and animals, including large forests of giant kelp; flourishing populations of fishes, invertebrates, and cetaceans; and diverse colonies of pinnipeds and marine birds. The sanctuary protects historic shipwrecks and Chumash Indian artifacts. In 2003 the California Fish and Game Commission established a network of twelve marine protected areas (MPAs) within the Channel Islands National Marine Sanctuary, a significant step forward for marine conservation. The network encompasses 146 square

nautical miles, or roughly 12 percent of sanctuary waters. Ten of the twelve MPAs are "no-take" marine reserves, which means no fishing is allowed. Two others are conservation areas that allow fishing for pelagic (open ocean) species and lobsters.

Sanctuary designated: September 22, 1980
Protected area: 1,658 square miles
Key species: California sea lion, elephant seal, harbor seal, blue whale, gray whale, dolphin, blue shark, brown pelican, western gull, abalone, garibaldi, and rockfish
Key habitats: Kelp forests, rocky shores, sandy beaches, seagrass meadows, deep rocky reefs, and open ocean

Sanctuary superintendent: Chris Mobley
Headquarters address:
113 Harbor Way, Ste. 150
Santa Barbara, CA 93109
ph: 805-966-7107
fax: 805-568-1582
channelislands@noaa.gov
http://channelislands.noaa.gov

Cordell Bank

Located at the edge of the continental shelf, sixty miles northwest of California's Golden Gate Bridge, Cordell Bank rises from the seafloor. Although the water around the bank is about 4,000 feet deep, this submerged island rises to within 120 feet of the ocean surface along a few of its ridges and pinnacles. The bank's topography creates upwellings of nutrient-rich ocean waters that provide a lush feeding ground for many marine mammals and seabirds. The depth, currents, and distance from the mainland have largely kept this special part of the California seafloor a mystery to both scientists and the public.

Sanctuary designated: May 24, 1989
Protected area: 526 square miles
Key species: Krill, Pacific salmon, rockfish, humpback whale, blue whale, Dall's porpoise, albatross, and shearwater
Key habitats: Rocky reefs, open ocean, soft sediment, and continental slope and shelf

Sanctuary manager: Dan Howard
Headquarters address:
One Bear Valley Rd.
Point Reyes Station, CA 95950
ph: 415-663-0314
fax: 415-663-0315
cordellbank@noaa.gov
http://cordellbank.noaa.gov

Gulf of the Farallones

Along the coast of California, north and west of San Francisco, the Gulf of the Farallones National Marine Sanctuary contains spawning grounds and nurseries for commercially valuable fish species, at least thirty-six species of marine mammals, and thirteen species of breeding seabirds. The Farallon Islands are home to the largest concentration of breeding seabirds in the contiguous United States, and one-fifth of California's harbor seals are thought to breed within the sanctuary. It is best known, however, for the dozens of great white sharks that come to feed in its waters every fall during elephant seal breeding season. The sanctuary boundaries include the coastline up to mean high tide, protecting a number of accessible lagoons, estuaries, bays, and beaches for the public.

Sanctuary designated: January 16, 1981
Protected area: 1,255 square miles
Key species: Steller sea lion, gray whale, blue whale, humpback whale, Dungeness crab, common murre, and ashy storm-petrel
Key habitats: Coastal beaches, rocky shores, salt marsh, estuaries, mudflats, tidal flats, open ocean, deep benthos, and continental slope and shelf

Sanctuary manager: Maria Brown
Headquarters address:
Fort Mason, Bldg. 201
San Francisco, CA 94123
ph: 415-561-6622
fax: 415-561-6616
farallones@noaa.gov
http://farallones.noaa.gov

Monterey Bay

Monterey Bay is the largest of the nation's thirteen marine sanctuaries, encompassing more than 5,300 square miles off central California. The sanctuary contains many diverse biological communities, from rugged rocky shores to lush kelp forests to one of the deepest underwater canyons in North America. An abundance of life, from tiny plankton to huge blue whales, thrives in these waters. This diversity of habitats and marine life has made the sanctuary a national focus for marine research and educational programs and a recreational destination for divers, boaters, surfers, and others.

Sanctuary designated: September 18, 1992
Protected area: 5,328 square miles along nearly 300 miles of the coast from the Marin County headlands south to Cambria
Key species: Sea otter, blue whale, market squid, brown pelican, rockfishes, giant kelp, krill, and leatherback sea turtle
Key habitats: Sandy beaches, rocky shores, kelp forests, subtidal rocky reefs, soft-bottom benthic submarine canyons, cold seeps, wetlands, and open ocean

Sanctuary superintendent: William J. Douros
Headquarters address:
299 Foam St.
Monterey, CA 93940
ph: 831-647-4201
fax: 831-647-4250
montereybay@noaa.gov
http://montereybay.noaa.gov

FLORIDA
Florida Keys

This sanctuary contains one of the most diverse underwater communities of plants and animals in North America. Its dazzling coral reefs support rich marine populations that depend on the reefs for shelter and food. This complex marine ecosystem, which also includes fringing mangroves, seagrass meadows, hard-bottom communities, and bank reefs, supports the commercial fishing and tourism-based businesses that are crucial to Florida's economy. The sanctuary also protects the final resting places of shipwrecks that span both precolonial and modern maritime history. Even though the sanctuary is involved in marine management and resource protection efforts, nutrient runoff, pollution, groundings, and coral bleaching increasingly threaten its reefs.

> Sanctuary designated: November 16, 1990
> Protected area: 3,674 square miles
> Key species: Hard corals (elkhorn, staghorn, pillar, brain, and star), soft corals (sea fans, sea rods, and sea whips), sponges, turtle grass, angelfish, spiny lobster, stone crab, grouper, and tarpon
> Key habitats: Coral reefs, patch and bank reefs, mangrove-fringed shorelines and islands, sand flats, seagrass meadows, hard-bottom communities, open ocean

Sanctuary superintendent: Billy Causey
Headquarters address:
PO Box 500368
Marathon, FL 33050
ph: 305-743-2437
fax: 305-743-2357
floridakeys@noaa.gov
http://floridakeys.noaa.gov
Regional operations offices are located in Key West and Key Largo.

GEORGIA
Gray's Reef

Gray's Reef lies seventeen miles east of Sapelo Island, Georgia, in waters fifty- to seventy-feet deep. One of the largest nearshore sandstone reefs in the southeastern United States, its sandy, flat-bottom troughs and sandstone outcroppings and ledges are optimal for the colonization of marine invertebrates such as sponges and soft corals. These inhabitants in turn attract a rich diversity of reef and pelagic fish, sea turtles, and marine mammals. The sanctuary also lies near the only known calving grounds for the northern right whale, the most endangered large whale in the world. In 1986 the United Nations designated Gray's Reef as an International Biosphere Reserve.

> Sanctuary designated: January 16, 1981
> Protected area: 23 square miles
> Key species: Loggerhead sea turtle, spotted and bottlenose dolphins, gag grouper, black sea bass, angelfish, barrel sponge, ivory bush coral, and sea whip
> Key habitats: Calcareous sandstone reefs, sandbottom communities, moderate relief ledges, patch reefs, and temperate reef

Sanctuary manager: Reed Bohne
Headquarters address:
10 Ocean Science Circle
Savannah, GA 31411
ph: 912-598-2345
fax: 912-598-2367
graysreef@noaa.gov
http://graysreef.noaa.gov

HAWAII
Hawaiian Islands Humpback Whale National Marine Sanctuary

One of the world's most important humpback whale habitats lies within the warm and shallow protected waters of the Hawaiian Islands Humpback Whale National Marine Sanctuary. Scientists estimate that two-thirds of the entire North Pacific humpback whale population migrates to Hawaiian waters to breed, calve, and nurse their newborns. The sanctuary is also home to a fascinating array of other marine animals, corals, and plants, some of which are found nowhere else on Earth. Its cultural heritage includes native Hawaiian traditions of living in harmony with the sea.

> Sanctuary designated: November 4, 1992
> Protected area: 1,370 square miles
> Key species: Humpback whale, pilot whale, Hawaiian monk seal, spinner dolphin, green sea turtle, coral reefs, and limu (seaweed)
> Key habitats: Humpback whale breeding, calving, and nursing grounds; coral reefs; and sandy beaches

Sanctuary manager: Naomi McIntosh
Headquarters address:
6700 Kalaniana'ole Hwy., #104
Honolulu, HI 96825
ph: 808-397-2651
fax: 808-397-2650
hihumpbackwhale@noaa.gov
http://hawaiihumpbackwhale.noaa.gov

Northwestern Hawaiian Islands Coral Reef Ecosystem Reserve

In recognition of the perilous state of the world's coral reefs, Executive Order 13178 established the Northwestern Hawaiian Islands Reserve in December 2000. Stretching nearly 1,500 miles, the reserve contains some of the healthiest and most undisturbed coral reefs on the planet and is one of the last predator-dominated coral reef ecosystems on Earth. Reserve waters are home to over 7,000 marine species, one-quarter of which are unique to the Hawaiian Archipelago. This uninhabited area contains critical habitats for many endangered and threatened species, including the Hawaiian monk seal and the green sea turtle. The reserve is the largest conservation area (marine or terrestrial) in the United States and the second largest in the world after Australia's Great Barrier Reef. The National Marine Sanctuary Program has begun the process of designating the reserve as America's fourteenth national marine sanctuary.

> Reserve established: December 4, 2000
> Protected area: 135,522 square miles
> Key species: More than 7,000 marine species, including the endangered Hawaiian monk seal and the threatened green sea turtle. One-fourth of all marine species are endemic and are found nowhere else on Earth.
> Key habitats: Atolls, coral reefs, seamounts, banks, shoals, and open ocean

Reserve coordinator: Robert Smith
Headquarters address:
308 Kamehameha Ave., Ste. 203
Hilo, HI 96720
ph: 808-933-8180
fax: 808-933-8186
hawaiireef@noaa.gov
http://hawaiireef.noaa.gov

MASSACHUSETTS
Stellwagen Bank

The slow retreat of massive Ice Age glaciers formed what is today Stellwagen Bank, a shallow sand and gravel plateau located at the mouth of Massachusetts Bay. Here, ocean currents sweep water in and out of the bay, with the bank funneling this flow into narrow passageways. The resulting nutrient-rich waters make this an area of high marine productivity, supporting a food web with species ranging from single-celled phytoplankton to great whales. This marine productivity supports local fishermen as well as a robust tourism industry. An estimated one million whale watchers visit the sanctuary waters each year, intent on witnessing the acrobatics of the gregarious but endangered humpback whale.

> Sanctuary designated: November 4, 1992
> Protected area: 842 square miles
> Key species: Humpback whale, northern right whale, white-sided dolphin, storm-petrel, northern gannet, bluefin tuna, Atlantic cod, winter flounder, sea scallop, and northern lobster
> Key habitats: Sand-and-gravel bank, muddy basins, boulder fields, rocky ledges, and open water

Sanctuary superintendent: Craig MacDonald
Headquarters address:
175 Edward Foster Rd.
Scituate, MA 02066
ph: 781-545-8026
fax: 781-545-8036
stellwagen@noaa.gov
http://stellwagen.noaa.gov

TEXAS
Flower Garden Banks

One hundred miles off the coasts of Texas and Louisiana, a trio of underwater gardens emerge from the depths of the Gulf of Mexico. These fertile coral reefs, the northernmost in the continental United States, serve as regional reservoirs of shallow water for Caribbean reef fishes and invertebrates. The salt domes of Flower Garden and Stetson banks provide homes for corals, sponges, and fish of such beauty that they have become a premier diving destination in the United States. Each winter, schools of hammerhead sharks and spotted eagle rays visit the sanctuary. In summer, coral spawning attracts scientists and divers from around the world.

> Sanctuary designated: January 17, 1992
> Protected area: 56 square miles
> Key species: Star coral, brain coral, manta ray, hammerhead shark, and loggerhead turtle
> Key habitats: Coral reefs, algal-sponge communities, brine seep, sand flats, and open ocean

Sanctuary manager: G. P. Schmahl
Headquarters address:
1200 Briarcrest Dr., Ste. 4000
Bryan, TX 77802
ph: 979-846-5942
fax: 979-846-5959
flowergarden@noaa.gov
http://flowergarden.noaa.gov

VIRGINIA
Monitor National Marine Sanctuary

Monitor National Marine Sanctuary enjoys the distinction of being the nation's first national marine sanctuary. The *Monitor* is recognized worldwide for its significance as the vessel that revolutionized nineteenth-century naval technology and warfare. In 1862, the turreted ironclad engaged the Confederate warship CSS *Virginia* in battle. Less than a year later, the *Monitor* sank in a storm off Cape Hatteras, North Carolina. Since 1995, the National Marine Sanctuary Program and its partners have worked to stabilize the *Monitor*'s hull and recover major components of the ship, including the propeller, propeller shaft, skeg, and steam engine. Divers recovered the *Monitor*'s famous revolving gun turret and cannon, along with the remains of two crewmen, in August 2002.

Sanctuary designated: January 30, 1975
Protected area: 1 square mile
Key species: Amberjack, black sea bass, red barbier, scad, dolphin, sand tiger shark, coral, sea anemone, and sea urchin
Key habitats: Open ocean, artificial reef

Sanctuary manager: John Broadwater
Headquarters address:
c/o The Mariners' Museum
100 Museum Dr.
Newport News, VA 23606
ph: 757-599-3122
fax: 757-591-7353
monitor@noaa.gov
http://monitor.noaa.gov

WASHINGTON
Olympic Coast

Spanning 3,310 square miles of marine waters and rugged beaches along Washington's Pacific coast, the Olympic Coast National Marine Sanctuary provides habitats for twenty-nine species of marine mammals and serves as a critical link along the Pacific flyway for migratory birds. The sanctuary protects a productive upwelling zone, home to rich marine mammal and seabird faunas, kelp forests, and invertebrate communities. The sanctuary is also home to over 150 documented shipwrecks and the vibrant contemporary cultures of the Quinault, Hoh, Quileute, and Makah nations. A variety of agencies monitor human activities such as shipping, noncommercial air traffic, and recreational uses to ensure preservation of this unique and largely undeveloped coastline.

Sanctuary designated: July 16, 1994
Protected area: 3,310 square miles, from Koitlah Point on Cape Flattery to the mouth of the Copalis River on Washington's Pacific coast
Key species: Tufted puffin, bald eagle, northern sea otter, gray whale, humpback whale, dolphin, Pacific salmon, and rockfish
Key habitats: Rocky and sandy shores, kelp forests, seastacks and islands, continental shelf, open ocean, and deepwater canyons

Sanctuary superintendent: Carol Bernthal
Headquarters address:
115 E. Railroad Ave., Ste. 301
Port Angeles, WA 98362-2925
ph: 360-457-6622
fax: 360-457-8496
olympiccoast@noaa.gov
http://olympiccoast.noaa.gov

NATIONAL PARKS AND SEASHORES

The U.S. national park system contains many outstanding examples of the nation's ocean heritage, both natural and cultural. More than 70 units of the system include 35 million acres of prime coastal habitats and 3 million acres of water along 4,800 miles of ocean shoreline. These ocean sites represent a collective multicultural history and commemorate numerous significant events and places in the nation's past. They provide common ground (and water) for recreation, understanding, and inspiration.[1]

For additional information on these blue-water sites, including maps and visitor guidance, go to the National Park Service Web site at www.NPS.gov/parks.

[1] Excerpted from *National Park Service Ocean Stewardship Strategy,* by Dr. Gary E. Davis. [Davis, Gary E. "Maintaining Unimpaired Ocean Resources and Experiences: A National Park Service Ocean Stewardship Strategy," *The George Wright Forum,* December 2004.]

ALASKA
Aniakchak National Monument and Preserve

The Aniakchak Caldera is the result of a series of eruptions, the latest in 1931. Nearly six miles in diameter and covering some ten square miles, it is one of the finest examples of dry caldera in the world. Located in the volcanically active Aleutian Mountains, the crater contains many outstanding examples of volcanic features, including lava flows, cinder cones, and explosion pits. The preserve also extends over seventy miles of coastline.

Bering Land Bridge National Preserve

The Bering Land Bridge National Preserve is one of the most remote national park areas, located on the Seward Peninsula in northwest Alaska. The preserve is a remnant of the land bridge that connected Asia with North America more than 13,000 years ago. The majority of this land bridge, once thousands of miles wide, now lies beneath the waters of the Chukchi and Bering seas.

Cape Kruesenstern National Monument

Cape Krusenstern in Alaska is a treeless coastal plain dotted with sizable lagoons, with gently rolling limestone hills in the background. Cape Krusenstern's bluffs and a series of 114 beach ridges record the changing shorelines of the Chukchi Sea over thousands of years.

Glacier Bay National Park and Preserve

The marine wilderness of Glacier Bay National Park and Preserve provides opportunities for adventure, a living laboratory for observing the ebb and flow of glaciers, and a chance to study life as it returns in the wake of retreating ice. Amidst majestic scenery, Glacier Bay offers a connection to a powerful and wild landscape.

Katmai National Park and Preserve

Katmai National Monument on the Alaska peninsula across from Kodiak Island preserves the famed Valley of Ten Thousand Smokes. At least fourteen volcanoes in Katmai are considered "active," but none of the volcanoes are currently erupting. Brown bears, numbering more than 2,000, and vast schools of salmon are very active in Katmai. During the peak of the world's largest sockeye salmon run each July, forty to sixty bears congregate in Brooks Camp. Along the 480-mile Katmai Coast, brown bears also enjoy clams, crabs, and an occasional whale carcass.

Kenai Fjords National Park

Sweeping from rocky coastline to glacier-crowned peaks, Kenai Fjords National Park encompasses 607,805 acres of unspoiled wilderness on the southeast coast of Alaska's Kenai Peninsula.

Klondike Gold Rush National Historical Park

This park celebrates the Klondike Gold Rush of 1897–98 through fifteen restored buildings in the Skagway Historic District on the town harbor in Southeast Alaska. The park also administers the Chilkoot Trail and a small portion of the White Pass Trail. Other trails lead to Smugglers Cove and the waters of Taiya Inlet.

Lake Clark National Park and Preserve

Lake Clark National Park and Preserve is a composite of ecosystems representative of many regions of Alaska. The spectacular scenery stretches from the shores of Cook Inlet across the Chigmit Mountains to the tundra-covered hills of the western interior.

Sitka National Historic Park

Established in 1910, Alaska's oldest federally designated park commemorates the 1804 Battle of Sitka. Located within this scenic 113-acre park, the Tlingit Fort and battlefield are all that remains of this last major conflict between Europeans and Alaska Natives. Southeast Alaska totem poles and a temperate rain forest setting combine to provide spectacular scenery along the park's coastal trail.

Wrangell-St. Elias National Park and Preserve

Alaska's Chugach, Wrangell, and St. Elias mountain ranges converge here in what is often referred to as the "mountain kingdom of North America." The largest unit of the National Park System, this spectacular park includes the continent's largest assemblage of glaciers and the greatest collection of peaks above 16,000 feet. Its coastal range is home to a number of marine mammals.

AMERICAN SAMOA
National Park of American Samoa

Two rain forest preserves and a coral reef are home to unique tropical animals, including the flying fox, Pacific boa, sea turtles, and an array of birds and fish. The park contains paleotropical rain forests, pristine coral reefs, and magnificent white sand beaches.

CALIFORNIA
Cabrillo National Monument

Cabrillo National Monument offers a superb view of San Diego's harbor and skyline. The Old Point Loma Lighthouse, a San Diego icon since 1854, stands at the park's highest point. In the winter, migrating gray whales can be seen off the coast.

Channel Islands National Park

Comprising five of a chain of eight southern California islands near Los Angeles, Channel Islands National Park is home to a wide variety of nationally and internationally significant natural and cultural resources. Over 2,000 species of plants and animals can be found within the park. Of these species, 145 are unique to the islands and found nowhere else in the world. Marine life ranges from microscopic plankton to the endangered blue whale, the largest animal to live on Earth.

Fort Point National Historic Site

The U.S. Army Corps of Engineers constructed Fort Point on the southern shore of the Golden Gate where the Pacific meets San Francisco Bay between 1853 and 1861. It was built to prevent the entry of a hostile fleet into the bay.

Golden Gate National Recreation Area

The Golden Gate National Recreation Area is one of the largest urban national parks in the world. It's 75,398 acres of land and water extend north of the Golden Gate Bridge to Tomales Bay in Marin County and south to San Mateo County, encompassing 59 miles of bay and ocean shoreline.

Point Reyes National Seashore

Point Reyes National Seashore contains unique elements of biological and historical interest in a spectacularly scenic panorama of thunderous ocean breakers, open grasslands, bushy hillsides, and forested ridges. Native land mammals number about thirty-seven species; marine mammals such as sea lions and elephant seals augment this total by another dozen species.

Presidio of San Francisco

Today, visitors enjoy the history and beauty of the former army base at the Presidio. The park encompasses more than 500 historic buildings, a collection of coastal defense fortifications, a national cemetery, a historic airfield, a saltwater marsh, forests, beaches, native plant habitats, coastal bluffs, miles of hiking and biking, and some of the most spectacular vistas in the world.

Redwood National Park

Redwood National and State parks are home to some of the world's tallest trees: old-growth coast redwoods. These trees can live to be 2,000 years old and grow to over 300 feet tall. The parks' mosaic of habitats include prairie/oak woodlands, mighty rivers and streams, and 37 miles of pristine Pacific coastline.

San Francisco Maritime National Historical Park

Located at the west end of San Francisco's Fisherman's Wharf, this park includes the fleet of national historic landmark vessels at the Hyde Street Pier, a visitor center, a maritime museum, and a library/research facility. Visitors can board turn-of-the-century ships, tour the museum, and learn traditional arts, such as boatbuilding and woodworking.

Santa Monica Mountains National Recreation Area

The Santa Monica Mountains rise above Los Angeles, widen to meet the curve of Santa Monica Bay, and reach their highest peaks facing the ocean, forming a beautiful and multifaceted landscape that (through its contiguous links to state parks) stretches from mountain peak to shore.

FLORIDA
Biscayne National Park

Biscayne Bay's shallow waters, south of Miami, may be glassy smooth—a window on another world—or may be covered with wind-driven white waves that bite like teeth at an angry sky. Some days, the water over the reef is so clear, making every detail on the bottom is visible. The largest marine park in the National Park System, 95 percent of Biscayne's 173,000 acres is covered by water. Its four primary ecosystems include the bay, mangrove forests, exposed keys, and underwater coral reefs.

Canaveral National Seashore

Located on a barrier island, Canaveral National Seashore includes ocean, beach, dune, hammock, lagoon, salt marsh, and pine flatland habitats. The park's north district is near New Smyrna Beach, Florida.

Castillo de San Marcos National Monument

Built during the period from 1672 to 1695, the Castillo de San Marcos served primarily as an outpost of the Spanish Empire, guarding St. Augustine, the first permanent European settlement in the continental United States. The Spanish remained in power in Florida until the United States purchased the area in 1821. The U.S. Army used the Castillo, renamed Fort Marion, as a coastal fort until 1899.

De Soto National Memorial

Visitors to the De Soto National Memorial can attend living history demonstrations, try on a piece of armor, or walk the nature trail through a Florida coastal landscape similar to the one conquistadors encountered almost 500 years ago.

Dry Tortugas National Park

First named The Turtles, *Las Tortugas*, by Ponce de Leon in 1513, these reefs and keys became known as the *Dry Tortugas* because they had no fresh water. In the nineteenth century the United States began construction of Fort Jefferson on Garden Key, one of the *Tortugas*. The fort later became a Union military prison during the Civil War. In 1992, the *Dry Tortugas*, seventy miles west of Key West, became a national park. Among its protected resources are endangered green sea turtles.

Everglades National Park

The Everglades National Park is a low, flat plain shaped by the action of water and weather. In the summer wet season it is a wide, grassy river. In the winter season the edge of the slough is a dry grassland. Along with its water marsh, alligators, turtles, and birds, the park encompasses extensive island chains and is home to marine animals in Florida Bay.

Fort Caroline National Memorial

Fort Caroline National Memorial memorializes the sixteenth-century French effort to establish a permanent colony in Florida. It is located on the water within the Timucuan Ecological and Historical Preserve.

Fort Matanzas National Monument

The Spanish built this outpost fort from 1740 to 1742 to guard the Matanzas Inlet and to warn the people of St. Augustine of British or other enemies approaching from the south. Fort Matanzas now serves as a reminder of the early Spanish empire in the New World. In addition, the park, which is located on barrier islands along the shores of the Atlantic Ocean and the Matanzas estuary, provides a natural habitat rich in wildlife.

Gulf Islands National Seashore

More than 80 percent of Gulf Islands National Seashore is under water, but the barrier islands are the most outstanding features to those who visit the area. The seashore stretches 160 miles from Cat Island in Mississippi to the eastern tip of Santa Rosa Island in Florida. Snowy-white beaches, sparkling blue waters, fertile coastal marshes, and dense maritime forests are among the many attractions.

Timucuan Ecological and Historical Preserve

The 46,000-acre Timucuan Ecological and Historic Preserve protects one of the last unspoiled coastal wetlands on the Atlantic Coast and preserves historic and prehistoric sites within the area. The estuarine ecosystem includes salt marsh, coastal dunes, and hardwood hammocks, all rich in native vegetation and animal life.

GEORGIA

Cumberland Island National Seashore

Cumberland Island is 17.5 miles long and totals 36,415 acres, nearly half of which are marsh, mudflats, and tidal creeks. It is well known for its sea turtles, wild turkeys, wild horses, armadillos, abundant shore birds, dune fields, maritime forests, salt marshes, and historic structures.

Fort Frederica National Monument

James Oglethorpe established Fort Frederica in 1736 to protect the southern boundary of his new colony of Georgia. After successfully repulsing a Spanish attempt to retake St. Simons Island, the garrison at Fort Frederica was disbanded in 1748, and the town fell into decline. Today, the National Park Service protects the archaeological remnants of Frederica, located on 240 acres of St. Simons Island off the coast of Georgia.

Fort Pulaski National Monument

Fort Pulaski's defining events, including its capture by Union forces in 1862, occurred during the American Civil War. Today, the park includes scenic marsh and uplands that support a variety of animal life characteristic of southern barrier islands.

GUAM

War in the Pacific National Historic Park

At War in the Pacific National Historical Park, the former battlefields, gun emplacements, trenches, and historic structures all serve as silent reminders of the bloody battles on the island of Guam during World War II. More than half the park, over 1,000 acres, is located offshore.

HAWAII

Ala Kahakai National Historical Trail

Established to preserve, protect, and interpret traditional native Hawaiian culture and natural resources, the Ala Kahakai Natural Historical Trail is a 175-mile trail corridor full of cultural and historical significance. Natural resources include nearshore reefs, estuarine ecosystems, coastal vegetation, migratory birds, a native sea turtle habitat, and several threatened and endangered species of plants and animals.

Haleakala National Park

This park preserves the outstanding volcanic landscape of the upper slopes of Haleakala on the island of Maui and protects the unique and fragile ecosystems of Kipahulu Valley, the scenic pools along Oheo Gulch, and many rare and endangered species.

Hawaii Volcanoes National Park

This park encompasses diverse environments that range from sea level to the summit of Mauna Loa, the earth's most massive volcano, at 13,677 feet. Kilauea, the world's most active volcano, offers scientists insights on the birth of the Hawaiian Islands and provides visitors with views of dramatic volcanic landscapes. More than half of the park is designated wilderness and provides unusual hiking and camping opportunities.

Kalaupapa National Historic Park

Two tragedies occurred on the Kalaupapa Peninsula on the north shore of the island of Moloka'i: the first was the removal of indigenous people in 1865 and 1895; the second was the forced isolation of lepers from 1866 until 1969. Kalaupapa National Historical Park contains the physical setting for these stories. Within its coastal boundaries are the historic Hansen's disease (leprosy) settlements and the churches associated with the work of Father Damien (Joseph De Veuster).

Kaloko-Honokohau National Historic Park

Kaloko-Honokohau, containing some 1,160 acres, is full of cultural and historical significance. Located on the west coast of the Big Island, it is the site of an ancient Hawaiian settlement that encompasses portions of four different *ahupua'a*, or traditional sea to mountain land divisions.

Pu'uhonua o Honaunau National Historic Park

Until the early nineteenth century, Hawaiians who broke a *kapu* (one of the ancient laws against the gods) could avoid certain death by fleeing to this beachfront place of refuge or *pu'uhonua*. The park, located on the west coast of the Big Island, now preserves this site.

Puukohola Heiau National Historic Site

The founding of the Hawaiian kingdom can be directly associated with one structure in the Hawaiian Islands: the temple known as Pu'ukohola Heiau located on the west coast of the Big Island. From 1790 to 1791 Kamehameha I (also known as Kamehameha the Great), together with chiefs and commoners, constructed the temple to incur the favor of the war god Kuka'ilimoku.

USS Arizona Memorial

Oil droplets bubble to the surface of Pearl Harbor above the USS *Arizona*, creating a vivid link to the past. On the quiet Sunday morning of December 7, 1941, a Japanese surprise air attack left the U.S. Pacific Fleet in smoldering heaps of broken, twisted steel. Within hours, 2,390 people lost their lives—half of them were casualties from the battleship *Arizona*.

LOUISIANA
Jean Lafitte National Historical Park and Preserve

Jean Lafitte National Historical Park and Preserve protects significant examples of the rich natural and cultural resources of Louisiana's Mississippi Delta region. The park consists of six physically separate sites and a park headquarters. The sites in Lafayette, Thibodaux, and Eunice interpret the Acadian culture of the area. The Barataria Preserve (in Marrero) interprets the natural and cultural history of the uplands, swamps, and marshlands of the region.

MAINE
Acadia National Park

Located on the rugged coast of Maine, Acadia National Park encompasses over 47,000 acres of granite-domed mountains, woodlands, lakes, ponds, and ocean shoreline. Such diverse habitats create striking scenery and make the park a haven for wildlife and plants.

MARYLAND
Assateague Island National Seashore

Storm-tossed seas as well as gentle breezes shape Assateague Island, which the states of Maryland and Virginia share. This barrier island is a tale of constant movement and change. Bands of wild horses freely roam among plants and native animals that have adapted to a life of sand, salt, and wind. Ghost crabs buried in the cool beach sand and tree swallows plucking bayberries on their southward migration offer glimpses of the animal world's connection to Assateague.

Fort McHenry National Monument and Historic Shrine

One thousand dedicated Americans in valiant defense of this fort, located on Baltimore Harbor, inspired Francis Scott Key to write "The Star-Spangled Banner." Regardless of the "rockets red glare, the bombs bursting in air," the defenders of Fort McHenry stopped the British advance on Baltimore and helped to preserve the United States of America.

MASSACHUSETTS
Boston Harbor Islands National Recreation Area

The Boston Harbor Islands area includes thirty-four islands situated within the Greater Boston shoreline. Rich in natural and cultural resources, the area invites visitors to explore tide pools; walk through a Civil War era fort; climb a lighthouse; hike lush trails and salt marshes; camp under the stars; or relax while fishing, picnicking, or swimming—all within reach of downtown Boston.

Cape Cod National Seashore

Cape Cod National Seashore comprises 43,604 acres of shoreline and upland landscape features, including a forty-mile-long stretch of pristine sandy beach. Its boundaries encompass a variety of historic structures, including lighthouses, a lifesaving station, and numerous Cape Cod–style houses. The seashore offers six swimming beaches, eleven self-guided nature trails, and a variety of picnic areas and scenic overlooks.

New Bedford Whaling National Historic Park

This park commemorates the heritage of the world's preeminent whaling port during the nineteenth century. A variety of cultural landscapes, historic buildings, museum collections, and archives preserve this history and recount the stories of a remarkable era.

Salem Maritime National Historic Site

Salem Maritime, the first National Historic Site in the National Park System, preserves and interprets the maritime history of New England and the United States. The site covers about nine acres of land and includes twelve historic structures along the waterfront in Salem, Massachusetts.

NEW YORK
Fire Island National Seashore

Pristine ocean shores, an ancient maritime forest, legacies of lighthouse keepers, and William Floyd's historic estate are just a few of the recreational, natural, and cultural resources of the Fire Island National Seashore. Located on New York's Long Island just one hour east of New York City, this park offers visitors many types of relaxation and educational opportunities.

Gateway National Recreation Area

Gateway National Recreation Area is a 26,000-acre recreation area in the heart of the New York metropolitan area. The park extends through three New York City boroughs, Brooklyn, Queens, and Staten Island, and into northern New Jersey. Park sites offer a variety of recreation opportunities, along with a chance to explore many significant cultural and natural resources, including the wetland salt marshes of Jamaica Bay and the waterfront beaches of the New York Bight.

NORTH CAROLINA
Cape Hatteras National Seashore

Stretching over seventy miles of barrier islands, Cape Hatteras National Seashore is a fascinating combination of natural and cultural resources and provides a wide variety of recreational opportunities. Once dubbed the "Graveyard of the Atlantic" for its treacherous currents, shoals, and storms, Cape Hatteras has a wealth of history relating to shipwrecks, lighthouses, and the U.S. Lifesaving Service. The park's fishing and surfing are considered the best on the East Coast.

Cape Lookout National Seashore

The seashore encompasses a fifty-six-mile-long section of the Outer Banks of North Carolina, running from Ocracoke Inlet on the northeast to Beaufort Inlet on the southeast. The three undeveloped barrier islands making up the seashore may seem barren and isolated, but they offer many natural and historical features that can make a visit rewarding.

Fort Raleigh National Historic Site

Just north of Manteo, near Cape Hatteras, this 450-acre coastal site commemorates the first English attempts at colonization in the New World (1585–87). The fate of this "lost colony" remains a mystery to this day. The park is also home to the outdoor symphonic drama *The Lost Colony*, performed each summer since 1937.

OREGON
Fort Clatsop National Memorial

This site, in the northwest corner of Oregon, commemorates the 1805–06 winter encampment of the thirty-three-member Lewis and Clark expedition. The park's focus is a replica of the explorers' fifty-foot-square Fort Clatsop. The memorial is being incorporated into a new Lewis and Clark National Historical Park that includes coastal state parks in Oregon and Washington commemorating the bicentennial of the expedition's arrival at and exploration of the Pacific shore.

SOUTH CAROLINA
Fort Sumter National Monument

America's most tragic conflict, the Civil War, ignited at Fort Sumter on April 12, 1861, when rebels attacked the Union fort. Located on a 200-acre island in the middle of Charleston Harbor, Fort Sumter is today managed by the National Park System and is accessible by ferry.

TEXAS
Padre Island National Seashore

Encompassing some 130,454 acres, Padre Island National Seashore is the longest remaining undeveloped stretch of barrier island in the world and offers a wide variety of flora and fauna as well as recreation. Five of the world's seven species of endangered sea turtles nest on the seashore. Controversies include ongoing and possible expansion of natural gas drilling within the park, even in sea turtle nesting areas.

VIRGINIA
Colonial National Historic Park

Colonial National Historical Park encompasses five sites spanning over 9,000 acres, including two of the most historically significant sites in English North America:

Jamestown, the first permanent English settlement in North America in 1607, and Yorktown Battlefield, the final major battle of the American Revolutionary War in 1781. These two sites represent the beginning and end of English colonial America. Located at the southern end of Chesapeake Bay, the park also contains a diversity of natural resources, including extensive wetlands. Part of the mid-Atlantic coastal plain, it has a direct hydrological link to the bay.

VIRGIN ISLANDS
Buck Island Reef National Monument

Buck Island Reef National Monument off St. Croix in the U.S. Virgin Islands preserves one of the finest marine gardens in the Caribbean Sea. The park is one of only a few fully marine-protected areas in the National Park System. The 176-acre island and surrounding coral reef ecosystem support a large variety of native flora and fauna, including endangered and threatened hawksbill turtles and brown pelicans. Both novice and expert snorkelers will enjoy the passages through the reef.

Christiansted National Historic Site

This seven-acre park centers on the Christiansted waterfront and wharf area on St. Croix and encompasses five historic structures on its grounds: Fort Christiansvaern (1738), the Danish West India and Guinea Company Warehouse (1749), the Steeple Building (1753), the Danish Custom House (1844), and the Scale House (1856).

Salt River Bay National Historic Park and Ecological Preserve

Salt River Bay is a living museum on St. Croix in the U.S. Virgin Islands. Prehistoric and colonial-era archaeological sites and ruins are found in a dynamic, tropical ecosystem that supports threatened and endangered species. The area's blend of sea and land holds some of the largest remaining mangrove forests in the Virgin Islands, as well as coral reefs and a submarine canyon.

Virgin Islands National Park

Renowned throughout the world for its breathtaking beauty, this park covers approximately three-fifths of St. John Island and nearly all of Hassel Island in the Charlotte Amalie harbor on St. Thomas. Within its borders lie protected bays of crystal blue-green waters teeming with coral reef life, white sandy beaches shaded by seagrape trees, coconut palms, and tropical forests providing habitats for over 800 species of plants.

WASHINGTON

Ebey's Landing National Historical Reserve
Located on Whidbey Island in Puget Sound, the 17,400-acre Ebey's Landing National Historical Reserve provides a vivid record of Pacific Northwest history, including the first exploration of Puget Sound by Captain George Vancouver in 1792. Among its features are the well-preserved nine-

teenth century seaport town of Coupeville on Penn Cove. The cove itself, with over 4,000 surface acres, provides a rich habitat for waterfowl and migratory birds. The western shore of the reserve includes eight miles of beaches and bluffs with a view of the Pacific across the rough waters of the Strait of Juan de Fuca.

Olympic National Park

This 922,000-acre park, covering the northwest corner of Washington state, embraces glacier-capped mountains, magnificent stands of old-growth forests, including temperate rain forests, and the wild Pacific coast. Its coastline is contiguous with the Olympic Coast National Marine Sanctuary.

San Juan Island National Historic Park

Orca whales and bald eagles abound here, as do more than 200 species of birds navigating the Pacific flyway. The 1,752-acre two-unit park is located on San Juan, one of the two largest islands in the San Juan Archipelago that includes over 800 islands, islets, rocks, and reefs. It is an excellent place to hike, picnic, play on the beach, experience wildlife, and enjoy a wealth of programming, including summer living history events.

NATIONAL WILDLIFE REFUGES ALONG THE COAST

Since President Teddy Roosevelt created the first National Wildlife Refuge at Pelican Island off Florida in 1903, this system of nature reserves, many of which allow hunting and fishing, has prospered under the U.S. Fish and Wildlife Service. Today there are more than 540 refuges on some 96 million acres of public lands. More than 160 of these refuges are located along America's shores and are accessible to the public.

For more information on visiting these coastal wonders go to www.refuges.fws.gov.

ALABAMA
Bon Secour National Wildlife Refuge

Bon Secour National Wildlife Refuge consists of 6,700 acres of wildlife habitat lying directly west of Gulf Shores and serves as a habitat for nongame migrating birds in the fall and spring. It was established in 1980.

ALASKA
Alaska Maritime National Wildlife Refuge

The Alaska Maritime National Wildlife Refuge is a place of great distances and great dramas, where winds whip through the grasses of rugged, wave-pounded islands and active volcanoes simmer, venting steam above collars of fog, while, nearby, great forests of kelp team with life. Because it is spread out along most of the 47,300 miles of Alaska's coastline, the sheer span of this 4.9-million acre refuge is difficult to grasp. More than 2,500 islands, islets, spires, rocks, reefs, waters, and headlands provide habitats for about 40 million seabirds or 80 percent of Alaska's nesting seabird population.

Alaska Peninsula National Wildlife Refuge

Sandwiched between the Becharof National Wildlife Refuge to the north and the Izembek National Wildlife Refuge to the south, the 3.7-million acre Alaska Peninsula National Wildlife Refuge presents a breathtakingly dramatic landscape made up of active volcanoes, towering mountain peaks, rolling tundra, and rugged wave-battered coastlines.

Arctic National Wildlife Refuge

Renowned for its wildlife, the Arctic National Wildlife Refuge covers much of northeastern Alaska, and its coastal plain on the Arctic Ocean is home to 45 species of land and marine mammals, including vast herds of caribou, 36 species of fish, and 180 species of birds. Plans by the George W. Bush administration to drill for oil in the ANWR have generated massive protests and environmental controversy.

Becharof National Wildlife Refuge

The 1.2-milliion acre Becharof National Wildlife Refuge by Bristol Bay presents a land of contrasts, from its rugged coastline to the 4,835-foot summit of the Mt. Peulik volcano. It includes everything from tundra and braided, glacier-fed rivers to sawtoothed mountain ranges, but few would argue that the biological heart of the refuge is the 35-mile by 15-mile lake that bears its name.

Izembek National Wildlife Refuge

Izembek National Wildlife Refuge at the tip of the Alaskan peninsula is the smallest (315,000 acres) and one of the most ecologically unique of Alaska's refuges, due in large measure to its extensive wetlands and eelgrass beds that provide vital habitats for marine wildlife and feed for migratory birds.

Kenai National Wildlife Refuge

Alaska's Kenai Peninsula's "youth" in geologic terms (glacial ice covered its entire landmass as recently as 10,000 years ago) does not take away from the wealth of habitats, scenery and wildlife in the Kenai National Wildlife Refuge, which draws a half a million visitors a year—more than any other wildlife refuge in Alaska.

Kodiak National Wildlife Refuge

Kodiak National Wildlife Refuge is a rugged, beautiful island on Alaska's southwestern coast. The refuge provides habitats for brown bear, salmon, and other wildlife, both marine and terrestrial. It was established in 1941.

Selawik National Wildlife Refuge

The Selawik National Wildlife Refuge is home to a variety of wildlife, including hundreds of thousands of caribou, moose, brown bear, and wolverine, while its approximately 24,000 lakes and wetlands also serve as breeding or stopover resting places for hundreds of thousands of migratory birds that feast on the many fish present in nearby waters.

Togiak National Wildlife Refuge

The Togiak National Wildlife Refuge is home to 48 mammal species (31 terrestrial and 17 marine) found at various times of year along the refuge's 600 miles of coastline. Staff and volunteers have sighted some 201 species of birds and have also documented more than 500 species of plants, demonstrating a high degree of biodiversity for a subarctic area.

Yukon Delta National Wildlife Refuge

The Yukon Delta National Wildlife Refuge encompasses approximately 22 million acres in the northern boreal zone of southwestern Alaska. The Yukon-Kuskokwim Delta is among the most populated rural areas in Alaska. Within the refuge, thirty-five villages and nearly 25,000 Yupik Eskimo people make their home.

AMERICAN SAMOA
Rose Atoll National Wildlife Refuge

The Rose Atoll National Wildlife Refuge, 14 degrees south of the equator and about 2,500 miles south of Hawaii, is the smallest atoll in the world. Its nearly 20 acres of land and 1,600 acres of lagoon form an important nesting area for the threatened green sea turtle and endangered hawksbill turtle.

CALIFORNIA
Castle Rock National Wildlife Refuge

Half a mile offshore from Crescent City in far northern California, this coastal rock covers approximately 14 acres and rises steeply to 335 feet above sea level. The refuge provides an important sanctuary for Aleutian Canada geese and nesting seabirds.

Don Edwards San Francisco Bay National Wildlife Refuge

The Don Edwards San Francisco Bay National Wildlife Refuge covers 30,000 acres in south San Francisco Bay and is one of the largest urban refuges in the United States—an island of wildlife habitats in a burgeoning metropolitan area of seven million people.

Farallon National Wildlife Refuge

Farallon's 211 acres of rocky islands are located approximately 30 miles offshore from San Francisco in the Pacific Ocean. The refuge has the largest seabird breeding colony on the Pacific Coast south of Alaska, hosting more than 300,000 birds each summer. Stellar sea lions and northern elephant seals breed and pup on the South Farallon Islands. Due to its sensitive nature, the Farallon refuge is not open to the public.

Guadalope-Nipomo Dunes National Wildlife Refuge

With the Pacific Ocean to the west and farmland to the east, the Guadalupe-Nipomo Dunes National Wildlife Refuge, located along California's central coast, encompasses one of the largest coastal dune systems remaining in California.

Humboldt Bay National Wildlife Refuge

The Humboldt Bay National Wildlife Refuge is located on Humboldt Bay on the northwest California coast, where it protects wetlands and bay habitats for migratory birds.

Marin Islands National Wildlife Refuge

The Marin Islands National Wildlife Refuge in Marin County, just north of San Francisco, covers 340 acres. Islands form the core of the refuge, with surrounding submerged tidelands. Sheltered coves and shallow mudflats support wintering populations of diving ducks and feeding sites for fledged herons and egrets.

Salinas River National Wildlife Refuge

The Salinas River National Wildlife Refuge on the central coast encompasses 367 acres and consists of several habitat types, including ocean, beach, grassland, river, sand dunes, pickleweed salt marsh, river lagoon, riverine, and a saline pond. The area provides habitats for several threatened and endangered species.

San Diego National Wildlife Refuge

The San Diego National Wildlife Refuge currently extends over approximately 9,478 acres. It offers protected, enhanced, and restored habitats for endangered species, migratory birds, and rare plants and animals found in a variety of habitats. It was established in 1997.

San Pablo National Wildlife Refuge

The San Pablo National Wildlife Refuge in the northern part of the San Francisco Bay/Delta complex encompasses open bay and tidal marshes, mudflats, and seasonal and managed wetland habitats. It provides critical migratory and wintering habitats for shorebirds and waterfowl as well as a year-round habitat for endangered, threatened, and sensitive species.

Seal Beach National Wildlife Refuge

The Seal Beach National Wildlife Refuge is located in southern coastal California and encompasses 911 acres of remnant saltwater marsh in the Anaheim Bay estuary. The refuge serves as a significant stopover and wintering area along the Pacific flyway for shorebirds.

Sweetwater Marsh National Wildlife Refuge

Sweetwater Marsh is an urban refuge located on San Diego Bay in Southern California. Comprising 316 acres of salt marsh and coastal uplands surrounded by urban development, it is a critically important area for wildlife because over 90 percent of the historic wetlands of San Diego Bay have been filled in, drained, or diked.

Tijuana Slough National Wildlife Refuge

The Tijuana Slough National Wildlife Refuge is located in the southwesternmost corner of the contiguous United States just below Imperial Beach by the U.S.–Mexico border. At just over 1,000 acres, it is one of southern California's largest remaining salt marshes without a road or railroad trestle running through it.

CONNECTICUT

Stewart B. McKinney National Wildlife Refuge

The McKinney National Wildlife Refuge comprises eight parcels of land that stretch across sixty miles of Connecticut's shoreline. It provides important resting, feeding, and nesting habitats for many species of wading birds, shorebirds, songbirds, and terns.

DELAWARE

Bombay Hook National Wildlife Refuge

Bombay Hook, in Kent County, hosts hundreds of thousands of migrating ducks, geese, shorebirds, and neotropical songbirds, all following traditions of natural history— the spring and fall migrations. Established in 1937, it covers 15,978 acres, four-fifths of which is a tidal salt marsh.

Prime Hook National Wildlife Refuge

The Prime Hook National Wildlife Refuge is near the western shore of Delaware Bay. It conserves an important segment of the Delaware Bay marshes to protect migrating and wintering waterfowl habitat. Prime Hook is considered to have one of the best existing wetland habitat areas along the Atlantic Coast. It was established in 1963.

FLORIDA

Archie Carr National Wildlife Refuge

The Archie Carr National Wildlife Refuge stretches 20.5 miles between Melbourne Beach and Wabasso Beach along Florida's eastern coast. It is the most important nesting area for endangered loggerhead sea turtles in the Western Hemisphere. Thirty-five percent of all green turtle nesting in the United States also occurs on these beaches. Densities of 1,000 nests per mile have been recorded. The refuge is named for the late Dr. Archie Carr Jr. in honor of his extraordinary contribution to sea turtle conservation. It was established in 1991.

Cedar Keys National Wildlife Refuge

Totaling 762 acres, the Cedar Keys National Wildlife Refuge on Florida's western coast comprises thirteen islands ranging in size from 1 acre to 120 acres. President Herbert Hoover established the refuge in 1929 to protect a breeding ground for colonial nesting migratory birds.

Chassahowitzka National Wildlife Refuge

The Chassahowitzka National Wildlife Refuge encompasses over 31,000 acres of saltwater bays, estuaries, and brackish marshes at the mouth of the Chassahowitzka River. Primarily, the refuge offers protected waterfowl habitats and is home to over 250 species of birds, over 50 species of reptiles and amphibians, and at least 25 species of mammals. It was established in 1941.

Crocodile Lake National Wildlife Refuge

Crocodile Lake protects critical breeding and nesting habitats for the endangered American crocodile and other wildlife. The refuge, located on Key Largo, currently covers 6,700 acres, including 650 acres of open water, and contains a mosaic of habitat types, such as tropical hardwood hammock, mangrove forest, and salt marsh. It was established in 1980.

Crystal River National Wildlife Refuge

The eighty-acre Crystal River National Wildlife Refuge on Florida's western coast, seventy miles north of St. Petersburg, offers protection for the endangered West Indian Manatee and preserves the warm water spring havens that provide a critical habitat for the migrating manatee populations each winter. It was established in 1983.

Egmont Key National Wildlife Refuge

The 320-acre Egmont Key located offshore from St. Petersburg protects a diverse community of animals and plants, many of which are either threatened or endangered. The site of the former U.S. Army Fort Dade Military Reservation, the island once quaked under the sound of massive coastal cannon fire but now resides quietly amid the sounds of wildlife as nature has taken over. It was established in 1974.

Great White Heron National Wildlife Refuge

The Great White Heron National Wildlife Refuge, located in the lower Florida Keys, consists of almost 200,000 acres of open water and islands. The islands, approximately 7,600 acres, are primarily mangroves. Some of the larger islands contain pine rockland and tropical hardwood hammock habitats that provide critical nesting, feeding, and resting areas for more than 250 species of birds. It was established in 1938.

Hobe Sound National Wildlife Refuge

Hobe Sound is a coastal refuge providing some of the most productive sea turtle nesting habitats in the United States. Its mainland tract consisting of sand pine scrub is valued because more than 90 percent of this community type has been lost to development in Florida. It was established in 1969.

Island Bay National Wildlife Refuge

The twenty-acre Island Bay National Wildlife Refuge, administered as a satellite of the J. N. "Ding" Darling refuge, is located in the Cape Haze area of Charlotte Harbor, Charlotte County, Florida, southwest of Punta Gorda, and consists of the higher portions of several islands and mangrove shorelines where shorebirds nest.

J. N. "Ding" Darling National Wildlife Refuge

The 6,354-acre J.N. "Ding" Darling National Wildlife Refuge is on the subtropical barrier island of Sanibel in the Gulf of Mexico. The refuge is part of the largest undeveloped mangrove ecosystem in the United States.

Key West National Wildlife Refuge

Established in 1908 by President Theodore Roosevelt as a preserve and breeding ground for native birds and other wildlife, the Key West National Wildlife Refuge is the first refuge established in the Florida Keys and one of the earliest refuges in the United States. Encompassing more than 200,000 acres (only 2,000 acres of land—the rest is water), the area is home to more than 250 species of birds, many varieties of marine wildlife, and an important sea turtle nesting area.

Lower Suwannee National Wildlife Refuge

The Lower Suwannee National Wildlife Refuge covers 54,000 acres and is one of the largest undeveloped river delta-estuarine systems in the United States. It protects natural ecosystems of the Suwannee River's lower reaches and coastal marsh as the river empties into the Gulf of Mexico. It was established in 1979.

Matlacha Pass National Wildlife Refuge

Matlacha Pass originally encompassed just three small islands that served as breeding grounds for native birds. Since then, the refuge, about 8 miles northwest of Ft. Myers, has grown to twenty-three islands encompassing

about 512 acres. Florida has designated the refuge as an aquatic preserve for threatened species including manatees, American crocodiles, and bald eagles. It was established in 1908.

Merritt Island National Wildlife Refuge

The Merritt Island National Wildlife Refuge, an overlay of the John F. Kennedy Space Center, provides a security buffer zone for NASA. Approximately one-half the refuge's 140,000 acres consist of brackish estuaries and marshes; the remaining lands consist of coastal dunes, scrub oaks, pine forests and flatwoods, and palm and oak hammocks. It was established in 1963.

National Key Deer National Wildlife Refuge

The National Key Deer National Wildlife Refuge protects and preserves Key deer and other wildlife resources in the Florida Keys. Currently, the refuge covers approximately 9,200 acres of land that includes pine rockland forests, tropical hardwood hammocks, freshwater wetlands, salt marsh wetlands, and mangrove forests. It was established in 1957.

Passage Key National Wildlife Refuge

The Passage Key National Wildlife Refuge is one of the first national wildlife refuges. This thirty-acre meandering barrier island hosts the largest royal tern and sandwich tern colonies in Florida. Because of its small size and importance to wildlife, the refuge is closed to all public use. It was established in 1905.

Pelican Island National Wildlife Refuge

Pelican Island holds a unique place in American history as the nation's first National Wildlife Refuge. Established to protect brown pelicans and other native birds nesting on the island, the refuge, located within the Indian River Lagoon, celebrated its centennial anniversary in 2003.

Pine Island National Wildlife Refuge

The Pine Island National Wildlife Refuge is on the southwestern coast of Florida, north of Sanibel Island in the Pine Island Sound. The 500-acre refuge encompasses more than seventeen islands and consists of densely forested red and black mangroves with little uplands habitat. It has also been designated as a Florida State Aquatic Preserve. It was established in 1908.

Pinellas National Wildlife Refuge

The Pinellas National Wildlife Refuge serves as a breeding ground for colonial bird species. One of the islands, Tarpon Key, hosts the largest brown pelican rookery in Florida. The abundant green seagrass beds around the island are protected from motorized boat activity. Because of its small size and importance to wildlife, the refuge is closed to all public use. It was established in 1951.

St. Marks National Wildlife Refuge

The 68,000-acre St. Marks National Wildlife Refuge, located twenty-five miles south of Tallahassee along the Gulf Coast of Florida, is a well-known oasis of natural Florida habitats for wildlife, especially birds. Natural salt marshes, freshwater swamps, pine forests, and lakes provide a haven for wildlife and people.

St. Vincent National Wildlife Refuge

St. Vincent is an undeveloped barrier island in the Gulf of Mexico just offshore from the mouth of the Apalachicola River. The area preserves its highly varied plant and animal communities in as natural a state as possible.

Ten Thousand Islands National Wildlife Refuge

The Ten Thousand Islands National Wildlife Refuge is in Collier County on the southwestern coast of Florida. Established in 1996, this 35,000-acre area protects important mangrove habitats and a rich diversity of native wildlife, including several endangered species like the American crocodile.

GEORGIA
Blackbeard Island National Wildlife Refuge

In 1940, the Blackbeard Island Reservation, located eighteen miles off the coast of Georgia, became the Blackbeard Island National Wildlife Refuge. The island consists of interconnecting linear dunes thickly covered by oak and palmetto vegetation. The refuge covers approximately 1,163 acres of open freshwater or freshwater marsh; 2,000 acres of regularly flooded salt marsh; 2,115 acres of maritime forest; and 340 acres of sandy beach.

Harris Neck National Wildlife Refuge

The Harris Neck National Wildlife Refuge in McIntosh County serves as an important link in the chain of refuges along the Atlantic seaboard and is the inland base for two neighboring barrier island refuges, Blackbeard Island and Wolf Island. Harris Neck contains notable concentrations of waterfowl, wading birds, and alligators.

Wassaw National Wildlife Refuge

Unlike many of Georgia's Golden Isles, little development and few management practices have modified Wassaw Island's primitive character. The 10,053-acre refuge includes beaches with rolling dunes, maritime forest, and vast salt marshes. It was established in 1969.

Wolf Island National Wildlife Refuge

The Wolf Island National Wildlife Refuge provides protection and habitat for migratory birds. Saltwater marshes cover more than three-quarters of the refuge's 5,126 acres. Wolf Island's saltwaters are open to a variety of recreational activities such as fishing and crabbing, but all beach, marsh, and upland areas are closed to the public. It was established in 1930 and designated as a National Wilderness Area in 1975.

GUAM
Guam National Wildlife Refuge

The Guam National Wildlife Refuge protects habitats, works to control nonnative species, and helps protect and recover endangered and threatened species. The refuge encompasses 771 acres (371 acres of coral reefs and 400 acres of terrestrial habitat). It was established in 1993.

HAWAII
Hawaiian Islands National Wildlife Refuge

This refuge consists of a chain of islands, reefs, and atolls extending about 800 miles northwest of the main Hawaiian Islands, for a total of 1,766 acres of emergent lands and 610,148 acres of submergent lands. It was established in 1909.

Huleia National Wildlife Refuge

Located on the southeast side of Kaua'i, Huleia National Wildlife Refuge covers approximately 241 acres that provide open, productive wetlands for endangered Hawaiian water birds. It was established in 1973.

Kakahaia National Wildlife Refuge

The Kakahaia National Wildlife Refuge, on the island of Moloka'i, is part of the Maui national wildlife refuge complex. It is home to two endangered species of seabirds, the Hawaiian stilt and Hawaiian coot. It was established in 1977.

Kilauea Point National Wildlife Refuge

Kilauea Point's rocky cliffs along the north shore of Kauai provide a premier nesting and roosting habitat for seven native Hawaiian seabirds, including the state bird, the nene. It is one of the most important seabird nesting sites in the inhabited Hawaiian Islands.

Pearl Harbor National Wildlife Refuge

This refuge consists of two units of 25 acres and 37 acres. They include ponds and wetlands that provide a habitat for endangered Hawaiian seabirds. Due to access difficulties, public use is restricted at both units and is prohibited during the stilt nesting season. It was established in 1976.

LOUISIANA
Bayou Sauvage National Wildlife Refuge

Located within the city limits of New Orleans, the Bayou Sauvage National Wildlife Refuge encompasses approximately 23,000 acres. It is the largest urban national wildlife refuge in the United States. It was authorized in 1986 and officially established in 1990.

Big Branch Marsh National Wildlife Refuge

The Big Branch Marsh National Wildlife Refuge comprises 15,000 acres of coastal marsh and pine forested wetlands. It protects some of the only Lake Pontchartrain shoreline existing in a natural state and provides habitats for a diversity of wildlife species. It was established in 1994.

Breton National Wildlife Refuge

The Breton National Wildlife Refuge stretches over a series of barrier islands in the Gulf of Mexico, whose sizes and shapes are constantly altered by tropical storms, wind, and tidal action. Now totaling 18,273 acres, the refuge provides sanctuary for nesting and wintering seabirds as well as a sandy beach habitat for a variety of wildlife species. It was established in 1904.

Delta National Wildlife Refuge

In 1862, approximately 100 miles below New Orleans, the Delta National Wildlife Refuge took shape when a breach in the natural levee of the Mississippi River occurred. Purchased by the federal government in 1935, the 48,000-acre refuge now provides sanctuary and a habitat to wintering waterfowl.

Sabine National Wildlife Refuge

The Sabine National Wildlife Refuge provides a habitat for migratory waterfowl and other birds. It encompasses 124,511 acres of fresh, intermediate, and brackish marshes and is one of the largest estuarine-dependent marine species nurseries in southwest Louisiana. It was established in 1937.

Shell Keys National Wildlife Refuge

Shell Keys is one of the oldest refuges in the national wildlife refuge system and is a testimony to the fast eroding shoreline of Louisiana. Due to its remote location on West Cote Blanche Bay, public access to the refuge is restricted. This Gulf of Mexico sand spit and barrier island provides a habitat for shorebirds, including brown pelicans and black skimmers. It was established in 1907.

MAINE

Cross Island National Wildlife Refuge

The Cross Island National Wildlife Refuge is a complex of six islands encompassing 1,700 acres in the town of Cutler where during the fall, thousands of waterfowl, songbirds, shorebirds, and raptors pass through on their southward migrations.

Franklin Island National Wildlife Refuge

Located in Muscongus Bay, this twelve-acre island supports a variety of bird species. The U.S. Coast Guard maintains the historic lighthouse on the island.

Moosehorn National Wildlife Refuge

Moosehorn is one of the northernmost national wildlife refuges in the Atlantic flyway. It provides an important feeding and nesting habitat for many bird species, including waterfowl, wading birds, shorebirds, upland game birds, songbirds, and birds of prey.

Petit Manan National Wildlife Refuge

The Petit Manan complex contains forty-five offshore islands and three mainland units. Totaling more than 7,400 acres and spanning over 200 miles of Maine coastline, the refuge offers protection to migratory birds. It was established between 1972 and 1980.

Pond Island National Wildlife Refuge

Pond Island National Wildlife Refuge is a ten-acre island located in the mouth of the Kennebec River. Its treeless character and grass, forb, and shrub cover provides an excellent habitat for nesting seabirds.

Rachel Carson National Wildlife Refuge

In cooperation with the state of Maine, the Rachel Carson National Wildlife Refuge protects valuable salt marshes and estuaries for migratory birds. It covers ten units and includes about 4,700 acres between Kittery and Cape Elizabeth, Maine. It was established in 1966.

Seal Island National Wildlife Refuge

In cooperation with the National Audubon Society, this sixty-five-acre island twenty-one miles off the coast of Rockland offers a protected habitat for colonial nesting seabirds.

MARYLAND
Blackwater National Wildlife Refuge

Located twelve miles south of Cambridge, Maryland, the Blackwater National Wildlife Refuge offers a haven for migratory waterfowl. It was established in 1933.

Eastern Neck National Wildlife Refuge

Located at the confluence of the Chester River and the Chesapeake Bay, the Eastern Neck National Wildlife Refuge is a 2,286-acre island sanctuary for migratory birds and provides a natural habitat for more than 240 bird species. It was established in 1962.

Martin National Wildlife Refuge

Martin National Wildlife Refuge includes 4,548 acres of tidal marsh, coves, and creeks on Smith Island in Chesapeake Bay. Its vegetated ridges form an important stopover and wintering area for thousands of migratory waterfowl and provide a nesting habitat for various wildlife species.

Susquehanna National Wildlife Refuge

The Susquehanna National Wildlife Refuge is a small island less than 1.5 acres in size. Located at the mouth of the Susquehanna River in Harford County, Maryland, it provides habitats for shorebirds, seabirds, and other wildlife.

MASSACHUSETTS
Mashpee National Wildlife Refuge

The Mashpee National Wildlife Refuge preserves and protects natural resources associated with the Waquoit Bay area to further protect waterfowl and wildlife. When complete, the refuge will total 5,871 acres. It was established in 1995.

Monomoy National Wildlife Refuge

Monomoy National Wildlife Refuge provides a haven for migratory birds with 7,604 acres of varied habitats such as oceans, salt and freshwater marshes, dunes, and freshwater ponds. It was established in 1944.

Nantucket National Wildlife Refuge

Encompassing twenty-four acres at Great Point, Nantucket National Wildlife Refuge is noted for its value in carrying out the national migratory bird management program. It was established in 1975.

Nomans Land Island National Wildlife Refuge

Surrounded entirely by the Atlantic Ocean, Nomans Land Island is a 628-acre refuge. Wetlands comprise about one-third of the island, ranging from emergent marshes to permanently flooded, open water set aside for the protection of migratory birds. The refuge is closed to all public uses.

Parker River National Wildlife Refuge

The Parker River National Wildlife Refuge occupies the southern three-fourths of Plum Island, an eight-mile barrier island, and provides feeding, resting, and nesting habitats for migratory birds. It was established in 1942.

Thacher Island National Wildlife Refuge

The Thacher Island National Wildlife Refuge provides feeding, resting, and nesting habitats for migratory birds. It occupies most of the northern half of the fifty-two-acre Thacher Island. It was established in 1972.

MISSISSIPPI
Grand Bay National Wildlife Refuge

The 14,000-acre Grand Bay National Wildlife Refuge protects one of the largest expanses of undisturbed pine savanna habitats in the Gulf Coastal Plain region. It was established in 1992 under the Emergency Wetlands Resources Act of 1986.

NEW HAMPSHIRE
Great Bay National Wildlife Refuge

Great Bay contains a variety of rich wildlife habitats from uplands to open waters. With its open coastal water and abundant prey, the refuge plays a significant role as migration and wintering habitat for the federally protected bald eagle. It was established in 1992.

NEW JERSEY
Cape May National Wildlife Refuge

The Cape May National Wildlife Refuge, at more than 11,000 acres, is strategically located to conserve the habitat for hundreds of thousands of migratory birds that pass through the area each year.

E. B. Forsythe National Wildlife Refuge

Formed from two distinct parks in 1984, the E. B. Forsythe complex covers approximately 46,000 acres along the southern New Jersey coast, of which nearly 80 percent is tidal salt meadow and marsh that is interspersed with shallow coves and bays. Most of the remainder is woodlands, with some fields that are maintained to provide habitat diversity.

Supawna Meadows National Wildlife Refuge

Currently, the Supawna Meadows National Wildlife Refuge owns approximately 2,800 acres within a 4,600-acre-approved boundary formed primarily of brackish tidal marshes that provide waterfowl with an important feeding and resting area, particularly during the fall and spring migrations. Supawna Meadows is a part of the Cape May national wildlife refuge complex.

NEW YORK
Amagansett National Wildlife Refuge

The Amagansett National Wildlife Refuge graces the shore of the Atlantic Ocean on Long Island's south fork. The thirty-six-acre refuge is of special significance in protecting and managing fragile shore habitat and wildlife. It was established in 1968.

Conscience Point National Wildlife Refuge

Conscience Point is a sixty-acre refuge on the north shore of Long Island's south fork that protects grasslands, oak-beech forest, shrub habitats, kettle holes, freshwater marshes, and salt marshes. It was established in 1971.

Elizabeth A. Morton National Wildlife Refuge

The 187-acre Elizabeth A. Morton National Wildlife Refuge on Long Island boasts exceptionally diverse habitats, including a bay beach, a brackish pond, a freshwater pond, kettle holes, tidal flats, a salt marsh, a freshwater marsh, shrub, grasslands, a maritime oak forest, and red cedar. It was established in 1954.

Oyster Bay National Wildlife Refuge

The Oyster Bay National Wildlife Refuge on the north shore of Long Island consists of over 3,000 acres of high-quality marine habitats that support a variety of aquatic-dependent wildlife.

Seatuck National Wildlife Refuge

Located on the south shore of Long Island, the Seatuck National Wildlife Refuge consists of 196 acres bordering the Great South Bay. Situated in a heavily developed urban area, the refuge is an oasis for many species of migratory birds and waterfowl.

Target Rock National Wildlife Refuge

Target Rock is on the north shore of Long Island, twenty-five miles east of New York City. The eighty-acre refuge encompasses mature oak-hickory forest, a half-mile rocky beach, a brackish pond, and several vernal ponds that all support a variety of songbirds, mammals, shorebirds, fish, reptiles, and amphibians.

Wertheim National Wildlife Refuge

The Wertheim National Wildlife Refuge on the south shore of Long Island is one of the last undeveloped estuary systems remaining on Long Island. Approximately half of the refuge consists of aquatic habitats; the refuge's salt marshes, combined with the adjacent New York State–owned salt marsh, form the largest continuous salt marsh on Long Island.

NORTH CAROLINA
Alligator River National Wildlife Refuge

The Alligator River National Wildlife Refuge, on the inside waters of the Outer Banks near Roanoke Island, contains 152,195 acres and preserves and protects a unique wetland habitat type—the pocosin—and its associated wildlife species. It was established in 1984.

Cedar Island National Wildlife Refuge

The Cedar Island National Wildlife Refuge consists of approximately 11,000 acres of irregularly flooded brackish marsh and 3,480 acres of pocosin and woodland habitats. It provides a wintering habitat for thousands of ducks and a nesting habitat for colonial water birds. It was established in 1964.

Currituck National Wildlife Refuge

The 4,000-acre Currituck National Wildlife Refuge preserves and protects the coastal barrier island ecosystem of Currituck County. It provides a wintering habitat for waterfowl and protects endangered species. It was established in 1984.

Mackay Island National Wildlife Refuge

Mackay Island's 8,100 acres provide a habitat for migratory waterfowl, primarily the greater snow goose. It is composed mostly of a marshy habitat along the Virginia–North Carolina border. It was established in 1960.

Pea Island National Wildlife Refuge

Pea Island's 5,834 acres are located on the northern end of Hatteras Island. It offers a refuge and breeding ground for migratory birds and other wildlife, including the greater snow goose and other migratory waterfowl. It was established in 1937.

Swanquarter National Wildlife Refuge

Swanquarter, on the northern shore of Pamlico Sound, consists of approximately 8,800 acres of salt marsh islands and forested wetland interspersed with potholes, creeks, and drains. It was established in 1932.

OREGON
Bandon Marsh National Wildlife Refuge

The 712-acre Bandon Marsh National Wildlife Refuge is located along the picturesque southern Oregon coast near the mouth of the Coquille River and protects the largest remaining tract of salt marshes within the Coquille River estuary.

Cape Meares National Wildlife Refuge

Cape Meares' 138 acres on Oregon's central coast protects one of the last remnants of coastal old-growth forest and provides a habitat for federally threatened bird species, including bald eagles and marbled murrelets. It was established in 1938.

Nestucca Bay National Wildlife Refuge

The verdant pastures around Nestucca Bay harbor six subspecies of Canada geese. In addition, the refuge includes wooded uplands, riparian wetlands, a salt marsh, and open meadows that provides habitats for waterfowl, shorebirds, mammals, amphibians, and much more.

Oregon Islands National Wildlife Refuge

From nearly every viewpoint on the Oregon coast, colossal rocks can be seen jutting out of the Pacific Ocean, creating postcard images. The Oregon Islands National Wildlife Refuge protects each of these rocks along the entire coast of Oregon. The wildlife found on offshore rocks, reefs, and islands is extremely susceptible to human disturbance, thus the islands are closed to public entry year-round.

Siletz Bay National Wildlife Refuge

The Siletz Bay National Wildlife Refuge, just south of Lincoln City, encompasses some of the most scenic estuarine habitats along the Oregon coastal highway, featuring a salt marsh, a brackish marsh, tidal sloughs, mudflats, and coniferous and deciduous forestland. The refuge provides nursery grounds for coho and chinook salmon and steelhead and cutthroat trout. Its primary ecological goal is to allow the salt marsh to return to its natural, tidally influenced state.

Three Arch Rocks National Wildlife Refuge
Designated as the first national wildlife refuge west of the Mississippi River, Three Arch Rocks lies a half-mile offshore of the community of Oceanside. One of the Oregon coast's best-known landmarks, the refuge consists of three large and six smaller rocks totaling fifteen acres. The refuge is one of the smallest designated wilderness areas in the country.

PACIFIC ISLANDS
Baker Island National Wildlife Refuge
Baker Island is a nearly level, saucer-shaped 405-acre island surrounded by a narrow reef and 30,504 acres of submerged land. Most of the refuge is a marine habitat, including extensive coral reefs and other inshore tropical ocean habitats.

Howland Island National Wildlife Refuge
Howland Island is a low, flat, sandy island with a narrow fringing reef. The refuge covers more than 32,000 acres, including the 400-acre Howland Island. The majority of the refuge is a marine habitat, including extensive coral reefs and other inshore tropical ocean habitats.

Jarvis Island National Wildlife Refuge
Jarvis Island is over 36,400 acres, including the 1,100-acre Jarvis Island that provides a nesting and roosting habitat for about twenty species of seabirds and shorebirds. The majority of the refuge is a marine habitat.

Johnston Atoll National Wildlife Refuge
The atoll comprises four small islands (696 acres), which constitute the only land area in more than 800,000 square miles of ocean. The refuge provides a critical, rat-free habitat for central Pacific sea bird populations, while its coral reef ecosystem is an important marine resource.

Kingman Reef National Wildlife Refuge
Besides a spectacular diversity of coral reef fishes, corals, and other marine organisms, Kingman Reef provides roosting, feeding, and other essential habitats for migratory seabirds and supports migratory shorebirds and threatened Pacific green turtles.

Midway Atoll National Wildlife Refuge
The Midway Atoll National Wildlife Refuge is a place of astonishing beauty that provides a nesting habitat for over two million sea birds. The waters surrounding Midway have been closed to commercial fishing for over sixty years, thus protecting more than 260 species of coral, fish, and other marine life.

Palmyra Atoll National Wildlife Refuge
Palmyra Atoll National Wildlife Refuge, a circular string of fifty-two islets, is one of the most diverse and spectacular coral reef systems in the world. More than 130 species of stony corals populate the reefs.

PUERTO RICO
Cabo Rojo National Wildlife Refuge
Cabo Rojo, on the southwestern side of Puerto Rico, is an upland buffer for the Cabo Rojo Salt Flats and offers a migratory bird habitat. In 1999 the Cabo Rojo Salt Flats were added to the refuge for a total of 1,836 acres. It was established in 1974.

Culebra National Wildlife Refuge
The 1,568-acre refuge comprises lands on the main island of Culebra and twenty-two smaller islands in the same vicinity and contains diverse habitats, including subtropical dry forest, mangroves, brush, and grasslands. Leatherback and hawksbill sea turtles use the refuge beaches for nesting.

Desecheo National Wildlife Refuge
Located fourteen miles west of Puerto Rico, the Atlantic Ocean and the Caribbean Sea form the boundaries of the island of Desecheo. The refuge encompasses the entire rugged 360-acre island. It is not open to the public because of safety considerations associated with unexploded ordnance that remain in an area that was previously used for military bombing practice. It is hoped that historic seabird colonies can be restored to the island.

Vieques National Wildlife Refuge
Over 17,500 acres, the Vieques National Wildlife Refuge contains several ecologically distinct habitats, including beaches, coastal lagoons, mangrove wetlands, and upland forested areas on both the eastern and western ends of the island. Some excellent examples of subtropical dry forest in the Caribbean can be found on refuge lands, including many sea turtle nesting beaches. Since the U.S. Navy stopped using the island for bombing practice, there is a new effort to protect and restore much of its native habitat. The refuge was established in 2003.

RHODE ISLAND
Block Island National Wildlife Refuge
Located approximately twelve miles offshore on picturesque Block Island, this 127-acre refuge provides an important habitat for wildlife and a place for people to appreciate its natural environment. It was established in 1973.

John H. Chafee National Wildlife Refuge
Located within the picturesque Narrow River on the southern coast of Rhode Island, this 317-acre refuge is comparatively small in size but big in protecting the unique features of this area, which includes a habitat for the largest black duck population in the state.

Ninigret National Wildlife Refuge

Perched on the shoreline of the largest saltpond in the state, the 900-acre Ninigret National Wildlife Refuge sits on the glacial outwash plain of the Charlestown moraine. Over 250 bird species visit the refuge and 70 nest there, including the threatened piping plover.

Sachuest Point National Wildlife Refuge

The 242-acre Sachuest Point National Wildlife Refuge is a popular site for over 65,000 annual visitors each year. It is renowned for its fantastic saltwater fishing and for the presence of the largest winter population of harlequin ducks on the East Coast.

Trustom Pond National Wildlife Refuge

Spanning 800 acres on the southern coast of Rhode Island, the Trustom Pond National Wildlife Refuge protects the only undeveloped salt pond in the state. Varied habitats in the refuge support over 300 species of birds, 40 species of mammals, and 20 species of reptiles and amphibians.

SOUTH CAROLINA
ACE Basin National Wildlife Refuge

At more than 11,000 acres, the ACE Basin National Wildlife Refuge, southwest of Charleston, helps protect the largest undeveloped estuary along the Atlantic Coast, with rich bottomland hardwoods and fresh and saltwater marshes offering food and cover to a variety of wildlife.

Cape Romain National Wildlife Refuge

Cape Romain provides wintering habitat for migratory birds. Its 64,000 acres encompass a twenty-mile segment of the Atlantic coast and include barrier islands, salt marshes, coastal waterways, fresh and brackish water impoundments, and maritime forest. It was established in 1932.

Pinckney Island National Wildlife Refuge

Pinckney Island is a 4,053-acre refuge that includes many islands—of which Pinckney Island is the largest and the only one open to public use. Nearly two-thirds of the refuge consists of salt marshes and tidal creeks. On Pinckney Island alone the land types include salt marsh, forestland, brushland, fallow field, and freshwater ponds—all supporting a diversity of bird and plant life. It was established in 1975.

Tybee National Wildlife Refuge

The Tybee National Wildlife Refuge is a breeding area for migratory birds and other wildlife covering 100 acres in the mouth of the Savannah River. It was established in 1938.

TEXAS
Anahuac National Wildlife Refuge

Located on the upper Texas Gulf Coast, this 34,000-acre haven for wildlife includes marshes combined with coastal prairie. Together, they provide a home for an abundance of wildlife, from migratory birds to alligators.

Aransas National Wildlife Refuge

Ringed by tidal marshes and broken by long, narrow sloughs, this 59,000-acre refuge sprawls mostly across the Blackjack Peninsula, where grasslands, live oaks, and red-bay thickets cover deep sandy soils. Storms and waters of the Gulf of Mexico constantly reshape this vital refuge that is home to over 390 different bird species.

Big Boggy National Wildlife Refuge

The Big Boggy National Wildlife Refuge near Matagorda consists of 5,000 acres of flat coastal prairies, salt marshes, and two large saltwater lakes that provide a habitat for migratory waterfowl and other bird species.

Brazoria National Wildlife Refuge

The protected coastal wetlands of the Brazoria National Wildlife Refuge on the Mid-Coast harbor more than 300 bird species where the expanse of salt and freshwater marshes, sloughs, ponds, coastal prairies, and bottomland forest represent feasting and lodging for all or part of the year. These vestiges of wild Texas offer exceptional wildlife watching for visitors.

Laguna Atascosa National Wildlife Refuge

World famous for its birds and home to a mix of wildlife found nowhere else, Laguna Atascosa National Wildlife Refuge is the largest protected area of natural habitats left in the Lower Rio Grande Valley. The refuge's 45,187 acres provide a valuable home to wildlife and to those who enjoy wildlife in wild lands. Herons, egrets, piping plovers, and roseate spoonbills can be found along the shores of the Lagoon Madre, as can American alligators.

Lower Rio Grande Valley National Wildlife Refuge

On the international border between the United States and Mexico, a host of nature's borders converge—climate, community, landforms, and geography. Only 5 percent of the native landscape remains on the lower river and its nearby reaches, yet the diversity within these fragments adds up to an astonishing 1,200 types of plants, 700 species of vertebrates (including nearly 500 bird species), and 300 kinds of butterflies. The refuge includes some eleven different biological communities covering over 90,000 acres, from the Chihuahuan thorn forest to tidal wetlands.

McFaddin National Wildlife Refuge

Located on the upper Texas Coast, the McFaddin National Wildlife Refuge provides an important feeding and resting habitat for migrating and wintering waterfowl populations. The 55,000-acre area covers the largest remaining freshwater marsh on the Texas Coast and thousands of acres of intermediate to brackish marsh. It was established in 1980.

Moody National Wildlife Refuge

The Moody National Wildlife Refuge is located on the Texas Gulf Coast. It was established in 1961.

San Bernard National Wildlife Refuge

Less than half of the 34,679-acre San Bernard National Wildlife Refuge is open to the public, leaving a vast landscape for a wildlife sanctuary. Yet a drive on the three-mile auto tour or hike on one of the three hiking trails can require a full day's worth of wildlife watching along the slough, salt marshes, and saltwater lakes.

Texas Point National Wildlife Refuge

Located on the upper Texas Coast, the Texas Point National Wildlife Refuge encompasses 8,900 acres of fresh to saltwater marsh, with some wooded uplands and prairie ridges. It was established in 1979.

VIRGINIA
Back Bay National Wildlife Refuge

The Back Bay National Wildlife Refuge, located near Virginia Beach, provides a habitat for migrating and wintering waterfowl. Because of its unique geographic location along the Atlantic Coast, the 8,500-acre area provides overlapping ranges and habitats for many birds and mammals, including snow geese, river otters, and, along the beach, ghost crabs, gulls, and shorebirds. It was established in 1938.

Chincoteague National Wildlife Refuge

Located primarily on the Virginia side of Assateague Island, the Chincoteague National Wildlife Refuge encompasses more than 14,000 acres of beach, dunes, marsh, and maritime forest, providing habitats for migratory birds and other species of wildlife and plants. It was established in 1943.

Eastern Shore of Virginia National Wildlife Refuge

The Eastern Shore of Virginia National Wildlife Refuge serves as one of the country's most valuable stopovers for migratory birds. Nestled between the Atlantic Ocean and Chesapeake Bay, this 1,127-acre area offers protection for migratory birds and endangered species. It was established in 1984.

Featherstone National Wildlife Refuge

Composed of wetlands and woodlands, the Featherstone National Wildlife Refuge is a narrow strip along the shore of the Potomac River and mouth of Neabsco Creek, providing a habitat for neotropical migrants, waterfowl, ospreys, and bald eagles. The refuge has no public access site and remains closed to public use.

Fisherman Island National Wildlife Refuge

The Fisherman Island National Wildlife Refuge is the southernmost barrier island, separated from the Eastern Shore of Virginia refuge by approximately one-half mile. Sand continues to expand the island's size, which is currently estimated at 1,850 acres. Because of the critical nature of its habitats for wildlife, including brown pelicans and royal terns, Fisherman Island is closed to the public.

Mason Neck National Wildlife Refuge

Eighteen miles south of Washington, D.C., the 2,276-acre Mason Neck National Wildlife Refuge protects a nesting, feeding, and roosting habitat for bald eagles. It is the first federal refuge established specifically for the (then endangered) bald eagle. It was established in 1969.

Occoquan National Wildlife Refuge

Situated at the confluence of the Potomac and Occoquan rivers, the 580-acre Occoquan Bay National Wildlife Refuge has a unique mix of wetlands, forest, and native grasslands that provides a diversity of habitats for a wide variety of species. It was established in 1998.

Plum Tree Island National Wildlife Refuge

Plum Tree Island consists of 3,450 acres of salt marsh, shrub-scrub, and wooded habitats that provide a haven for waterfowl, marsh birds, and shorebirds. Previously owned by the U.S. Department of Defense, the refuge is closed to the public for all purposes due to hazards of unexploded ordinance.

Wallops Island National Wildlife Refuge

The refuge, comprising mainly salt marshes and woodlands, contains habitats for a variety of trust species, including upland- and wetland-dependent migratory birds. Along with 375 acres in the refuge, the U.S. Fish and Wildlife Service has agreed to manage an adjacent 3,000 acres owned by NASA. Much of this area is salt marsh and includes sea-level fens and rare nutrient-poor maritime seepage wetlands.

VIRGIN ISLANDS
Buck Island National Wildlife Refuge

The U.S. Fish and Wildlife Service acquired the 45-acre Buck Island because of its value for migratory birds. Although the rocky coastline offers little in the way of recreation potential, the surrounding waters contain reefs and a shipwreck that attract large numbers of snorkelers, divers, and boaters.

Green Cay National Wildlife Refuge

The Green Cay National Wildlife Refuge is a small island located off the northern coast of St. Croix and consists of dry, forested areas and small cobble beaches. Now a part of the national wildlife refuge system, this fourteen-acre island would have been the first land that Christopher Columbus saw in 1493. It was established in 1977.

Sandy Point National Wildlife Refuge

Located at the southwest end of St. Croix, the Sandy Point National Wildlife Refuge consists of more than 360 acres of subtropical dry vegetation that includes the largest salt pond in the Virgin Islands and a continuous stretch of 3.2 kilometers of sandy beach.

Copalis National Wildlife Refuge

The Copalis National Wildlife Refuge consists of a portion of 870 islands, rocks, and reefs extending for more than 100 miles along Washington's Pacific Coast from Cape Flattery to Copalis Beach. These islands are a vital sanctuary where fourteen species of seabirds nest and raise their young.

Dungeness National Wildlife Refuge

Located along the northern coast of the Olympic Peninsula in Washington, the 630-acre Dungeness National Wildlife Refuge is home to one of the world's longest natural sand spits, which softens the rough sea waves to form a quiet bay and harbor, gravel beaches, and tide flats.

Flattery Rocks National Wildlife Refuge

Flattery Rocks consists of portions of 870 islands, rocks, and reefs that extends over 100 miles of coastline. It is a vital sanctuary where fourteen species of seabirds nest and raise their young. Sea lions, harbor seals, sea otters, and whales may also be seen around the islands.

Grays Harbor National Wildlife Refuge

The muddy tidal flats of Grays Harbor Estuary are one of four major staging areas for shorebirds in North America and home to one of the largest concentrations of shorebirds on the West Coast south of Alaska. The 1,500-acre refuge is located near Hoquiam, Washington.

Lewis and Clark National Wildlife Refuge

The Lewis and Clark National Wildlife Refuge near the mouth of the Columbia River hosts thousands of waterbirds, such as great blue herons, gulls, and shorebirds, that feed in the shallows and mudflats. It contains some twenty named islands totaling about 8,300 acres, as well as about 35,000 acres of sandbars, tidal marshes, mudflats, and open water.

Nisqually National Wildlife Refuge

The Nisqually National Wildlife Refuge is located where the fresh water of the Nisqually River meets the salt water of South Puget Sound, creating the Nisqually River Delta—a biologically rich and diverse area that supports a variety of habitats, including the estuary, freshwater wetlands, and riparian woodlands. It was established in 1974.

Protection Island National Wildlife Refuge

Located near the mouth of Discovery Bay in the Strait of Juan de Fuca, the 364-acre Protection Island National Wildlife Refuge holds the distinction of hosting 70 percent of the nesting seabird population of Puget Sound and the Strait of Juan de Fuca. Along with a nesting colony of tufted puffins, the island also provides resting and pupping areas for about 1,000 harbor seals.

San Juan Islands National Wildlife Refuge

The San Juan Islands National Wildlife Refuge includes reefs, grassy and forested islands, and eighty-three rocks scattered throughout the San Juan Islands of North Puget Sound. These islands, totaling almost 450 acres, protect colonies of nesting seabirds.

Willapa National Wildlife Refuge

Located on the shores of Willapa Bay near the Pacific Ocean, this 14,000-acre refuge boasts of one of the most pristine estuaries in the United States. The shallow water and mudflats of the bay support vast beds of eelgrass and shellfish and provide a spawning habitat for fish.

ESTUARINE RESEARCH RESERVES

The National Estuarine Research Reserve System is a network of twenty-six protected areas, all but one of which are on a coast. Established for long-term research, education, and stewardship, these areas include wetland bays, lagoons, and sloughs that act as the spawning grounds and nurseries of the sea. This partnership program between the National Oceanic and Atmospheric Administration (NOAA) and various states protects more than one million acres of estuarine land and water, offering educational opportunities to students, teachers, and the public. For information on visiting a reserve near you go to www.nerrs.noaa.gov.

ALABAMA
Weeks Bay Estuarine Research Reserve

Weeks Bay is a small estuary (about three square miles) receiving fresh water from the Magnolia and Fish rivers and draining a 198-square-mile watershed into a portion of Mobile Bay. It includes over 6,000 acres of coastal wetlands and water bottoms that provide rich and diverse habitats for a variety of fish, crustaceans, and shellfish, as well as many unique and rare plants.

ALASKA
Kachemak Bay Estuarine Research Reserve

Located on the western coast of the Kenai Peninsula, this 365,000-acre reserve is not only the largest in the system but also is one of the most productive, diverse, and intensively used estuaries in Alaska. The local community pursued the designation of Kachemak Bay as a National Estuarine Research Reserve.

CALIFORNIA
Elkhorn Slough Estuarine Research Reserve
Among the relatively few coastal wetlands remaining in California, the main channel of this slough winds inland nearly seven miles and is flanked by a broad salt marsh second in size in California only to San Francisco Bay. The reserve lands include oak woodlands, grasslands, and freshwater ponds that provide essential coastal habitats supporting a great diversity of native organisms and migratory animals.

San Francisco Estuarine Research Reserve
San Francisco Bay has lost nearly 97 percent of its historic tidal wetlands due to development pressures within and around the bay. Since 1999, approximately 11,420 acres of wetlands have been restored to tidal influence in San Francisco Bay. Plans worked out between the government, private industry, and environmentalists for restoring an additional 25,500 acres are underway. Tidal wetlands are critical for flood prevention; sediment management; and as a habitat for small mammals, migratory birds, and fish, many of which are threatened and endangered.

DELAWARE
Delaware Estuarine Research Reserve
Just under 5,000 acres, the Delaware Estuarine Research Reserve consists of two unique components: one in which the St. Joseph River freshwater wetlands, ponds, and forest lands dominate; another where the salt marsh and open water habitats of the Delaware Bay dominate.

FLORIDA
Apalachicola Estuarine Research Reserve
Apalachicola Bay is one of the most productive estuarine systems in the Northern Hemisphere, with 60 to 85 percent of the local population making their living off the fishing industry. The 246,000-acre reserve, both aquatic and onshore, protects the region's biological diversity as well as the economic value of its natural resources.

Guana Tolomato Matanzas Estuarine Research Reserve
The Guana Tolomato Matanzas Estuarine Research Reserve encompasses approximately 55,000 acres of salt marsh and mangrove tidal wetlands, oyster bars, estuarine lagoons, upland habitat, and offshore seas in northeastern Florida. It contains the northernmost extent of mangrove habitat on the East Coast of the United States.

Rookery Bay Estuarine Research Reserve
Located at the northern end of the Ten Thousand Islands on Florida's Gulf Coast, the Rookery Bay Estuarine Research Reserve represents one of the few remaining undisturbed mangrove estuaries in North America. It is a prime example of a nearly pristine subtropical mangrove forested estuary.

GEORGIA
Sapelo Island Estuarine Research Reserve
At 17,000 acres, Sapelo Island is the fourth largest Georgia barrier island and one of the most pristine. The 6,000-acre reserve consists of salt marshes, maritime forests, and beach dune areas. The island rich in natural and human history and dating back 4,000 years includes both native peoples and present day Gullah, descendents of African slaves who retained some of their original language and culture and, following the Civil War, stayed on to establish fishing and crabbing communities.

MAINE
Wells Estuarine Research Reserve
The 1,600-acre Wells Estuarine Research Reserve protects fields, forests, freshwater wetlands, salt marshes, and sandy beaches on the densely populated southern coast of Maine. Its diverse habitats support a broad variety of plants and animals, including rare species such as least terns, piping plovers, and slender blue flag iris.

MARYLAND
Chesapeake Bay Estuarine Research Reserve
The Chesapeake Bay is the largest estuary in the United States and was once one of the most productive bodies of water in the world for fish, oysters, and crabs. The reserve manages three protected estuarine areas, totaling over 10,000 acres of natural field laboratories, and seeks to develop and implement a coordinated program of research, monitoring, education, and volunteer activities in habitats including salt marshes, freshwater marshes, and a tidal riverine system.

MASSACHUSETTS
Waquoit Bay Estuarine Research Reserve
The 2,600-acre Waquoit Bay Estuarine Research Reserve encompasses open waters, barrier beaches, marshlands, and uplands on the southern shore of Cape Cod. It is representative of the northern section (Cape Cod to Sandy Hook) of the Virginian biogeographic region.

MISSISSIPPI
Grand Bay Estuarine Research Reserve
The 18,400-acre Grand Bay Estuarine Research Reserve in Jackson County is one of the most biologically productive estuarine ecosystems in the Gulf of Mexico region, supporting several rare or endangered plant and animal species, numerous important marine fishery resources, diverse habitat types, and archaeological sites.

NEW HAMPSHIRE
Great Bay Estuarine Research Reserve
The 5,280-acre Great Bay Estuarine Research Reserve offers a diversity of land and water areas around the Great Bay estuary, such as salt marshes, rocky shores, bluffs, woodlands, open fields, riverine systems, and tidal waters.

NEW JERSEY
Jacques Cousteau Estuarine Research Reserve
The Jacques Cousteau Estuarine Research Reserve encompasses over 114,000 acres in southeastern New Jersey, including a great variety of terrestrial, wetland, and aquatic habitats within the Mullica River–Great Bay ecosystem. It is regarded as one of the least disturbed estuaries in the densely populated urban corridor of the northeastern United States.

NEW YORK
Hudson River Estuarine Research Reserve
The Hudson River is an amazing resource: a tidal river that links the communities of the valley economically, culturally, and ecologically and a cradle of human development for thousands of years. The reserve consists of a series of wetlands (totaling less than 5,000 acres) located along 100 miles of the river most impacted by ocean tides.

NORTH CAROLINA
North Carolina Estuarine Research Reserve
North Carolina's estuarine system is the third largest in the country, encompassing more the two million acres. The North Carolina Estuarine Research Reserve preserves these fragile natural areas and the variety of life they support.

OREGON
South Slough Estuarine Research Reserve
The 4,779-acre South Slough Estuarine Research Reserve contains upland forests, freshwater wetlands and ponds, salt marshes, mudflats, eelgrass meadows, and open water habitats. South Slough is the southwestern arm of the larger Coos estuary on the southern coast of Oregon. The Coos estuary is an example of a drowned river mouth estuary.

PUERTO RICO
Jobos Bay Estuarine Research Reserve
The 2,883-acre Jobos Bay reserve on the southern coast of Puerto Rico is the second largest estuarine area in Puerto Rico. It encompasses a chain of fifteen tear-shaped mangrove islets known as Cayos Caribe and the Mar Negro area in western Jobos Bay. The reserve is home to the endangered brown pelican, the peregrine falcon, the hawksbill sea turtle, and the West Indian manatee.

RHODE ISLAND
Narragansett Bay Estuarine Research Reserve
The Narragansett Bay reserve encompasses 2,353 acres of land on Prudence, Patience, and Hope islands and 1,591 acres of water adjoining the islands out to a depth of eighteen feet. Prudence Island supports one of the densest herds of white tailed deer in the Northeast. Because Prudence Island is just a short distance from the entirely wooded Patience Island, deer can swim between the islands.

SOUTH CAROLINA
ACE Basin Estuarine Research Reserve
The 134,710-acre ACE Basin Estuarine Research Reserve protects the natural beauty and abundant wildlife of the area located along the state's southern coast. In addition, the reserve preserves habitats for many endangered or threatened species, such as the shortnosed sturgeon, loggerhead turtles, and bald eagles.

North Inlet–Winyah Estuarine Research Reserve
The 12,327-acre reserve features the salt marshes and ocean-dominated tidal creeks of the North Inlet Estuary plus the brackish waters and marshes of the adjacent Winyah Bay Estuary. The reserve is home to many threatened and endangered species, including sea turtles, sturgeons, least terns, and wood storks.

VIRGINIA
Chesapeake Bay Estuarine Research Reserve
As the nation's largest estuary, the Chesapeake Bay contains a diverse collection of habitats, including oyster reefs, seagrass beds, tidal wetlands, sandy shoals, and mudflats. To address the diversity of habitats, the Chesapeake Bay reserve established a 4,435-acre multisite system from tidal freshwater to high salinity conditions along the York River estuary.

WASHINGTON
Padilla Bay Estuarine Research Reserve
Padilla Bay, located in the Salish Sea, is part of a large river delta in a fjord estuary, carved out by glaciers that retreated about 20,000 years ago. Most of the 11,000-acre reserve in far northern Washington encompasses extensive seagrass meadows, tidal flats and sloughs, salt marshes, and upland forests and meadows.

GEOGRAPHICAL INDEX

District of Columbia

Florida

U.S. Food and Drug Administration (FDA), 110
U.S. Naval Academy, 62, 128
University of Maryland
    Center for Environmental Science, 128
    Maryland Sea Grant College, 136

Massachusetts

Association for the Preservation of Cape Cod, 62
Audubon Society, 62
Boston Harbor Association, 62
Boston Harbor Islands National Recreation Area, 147
Boston University Marine Program, 128
Buzzards Baykeeper/Coalition for Buzzards Bay, 62
Cape Cod Center for Sustainability, 62
Cape Cod Commercial Hook Fisherman's Association, 62
Cape Cod Commission, 62
Cape Cod National Seashore, 147
Cape Cod Stranding Network, 63
Center for Coastal Studies, 63
Center for Oceanic Research and Education, 63
Charles River Conservancy, 63
Charles River Watershed Association, 63
Citizens for the Protection of Waquoit Bay, 63
Coastal Management Program, 106
Conservation Law Foundation, 63
Department of Fisheries, Wildlife and Environmental Law
    Enforcement (Division of Fisheries and Wildlife), 115
Earthwatch Institute, 63
Environmental League, 63
estuarine research reserve, 161
Friends of Pleasant Bay, 63
Gloucester Fishermen's Wives Association, 64
Gulf of Maine Council on the Marine Environment
    Massachusetts Coastal Zone Management, 64
Hands Across the River Coalition, 64
International Fund for Animal Welfare, 64
Ipswich River Watershed Association, 64
Island Alliance, 64
IWC and the Whale Adoption Project, 64
Jason Foundation for Education, 64
Manomet Center for Conservation Sciences, 64
Marine Biological Laboratory, 128
Marine Education Center of Cape Ann, 64
Mashpee Environmental Coalition, 65
Mashpee National Wildlife Refuge, 155
Massachusetts Bay Marine Studies Consortium, 65
Massachusetts Bays National Estuaries Program, 65
Massachusetts Environmental Trust, 65
Massachusetts Institute of Technology
    MIT Sea Grant, 136
    MIT/WHOI Joint Program in Oceanography, 129
Massachusetts Land Trust Coalition, 65
Massachusetts Maritime Academy, 129
Massachusetts PIRG, 65
Massachusetts Society of Conservation Biology, 65
Massachusetts Water Watch, 65
Massachusetts Wildlife Federation, 65
Merrimack River Watershed Council, 66
Monomoy National Wildlife Refuge, 155
Nantucket National Wildlife Refuge, 155
National Environmental Law Center, 66
national historic sites, 147
national historical parks and preserves, 147
national marine sanctuary, 67, 142
national recreation areas, 147
national seashores, 147
national wildlife refuges, 155

The Nature Conservancy, 66
New Bedford Whaling Museum, 66
New Bedford Whaling National Historic Park, 147
New England Aquarium, 66
New England Campaign for a Healthy Ocean, 66
New England Fishery Management Council, 108
Nomans Land Island National Wildlife Refuge, 155
Northeastern University (Marine Science Center), 128
Ocean Alliance, 66
Office of Coastal Zone Management, 106
Parker River Clean Water Association, 66
Parker River National Wildlife Refuge, 155
Salem Maritime National Historic Site, 147
Salem Sound Coastwatch, 66
Save the Harbor/Save the Bay, 67
Sea Education Association, 67
Sea Grant Consortiums, 136
Sierra Club, 67
Stellwagen Bank National Marine Sanctuary, 67, 142
Surfrider Foundation, 67
SWIM (Safer Waters in MA), 67
Thacher Island National Wildlife Refuge, 155
Three Bays Preservation Society, 67
Trout Unlimited, 67–68
University of Massachusetts (Intercampus Graduate School of
    Marine Sciences and Technology), 128
Urban Ecology Institute, 68
Waquoit Bay Estuarine Research Reserve, 161
Whale Conservation Institute, 68
Woods Hole Oceanographic Institution, 68
    Marine Policy Center, 128
    WHOI Sea Grant Program, 136

Minnesota

Institute for Agriculture and Trade Policy, 68
Mississippi River Basin Alliance, 68

Mississippi

Audubon Society, 68
Coastal Management Program, 106
Department of Marine Resources, 115
    Office of Coastal Ecology, 106
Department of Wildlife, Fisheries, and Parks, 115
estuarine research reserve, 161
Grand Bay Estuarine Research Reserve, 161
Grand Bay National Wildlife Refuge, 155
Gulf Coast Research Laboratory, 129
Gulf States Marine Fisheries Commission, 109
Mississippi Wildlife Federation, 68
Mississippi–Alabama Sea Grant Consortium, 68
MS-AL Sea Grant Consortium, 137
National Marine Educators Association, 69
national wildlife refuge, 155
The Nature Conservancy, 69
Sea Grant Consortium, 137
Sierra Club, 69
University of Southern Mississippi (Department of Marine
    Science), 129

New Hampshire

Audubon Society, 69
Blue Ocean Society for Marine Conservation, 69
Clean Water Action/Clean Water Fund, 69
Coastal Management Program, 106
estuarine research reserve, 162
Fish and Game Department, 115
Great Bay Coast Watch, 69

Lower Laguna Madre Foundation, 91
Lower Rio Grande Valley National Wildlife Refuge, 158
McFaddin National Wildlife Refuge, 158
Moody National Wildlife Refuge, 158
national marine sanctuaries, 89, 142
national wildlife refuges, 158–159
Padre Island Field Research Station
    (U.S. Geological Survey), 133
Padre Island National Seashore, 148
Parks and Wildlife Department, 115
Rio Grande Delta Audubon, 91
San Bernard National Wildlife Refuge, 159
Sea Grant Consortium, 137
Shoreline Environmental Research Facility, 134
Sierra Club, 91
Surfrider Foundation, 92
Texas A&M University
    Center for Coastal Studies, 133
    Department of Marine Science (Laboratory for Oceanographic
        and Environmental Research), 133
    Department of Oceanography, 133
Texas Institute of Oceanography, 134
Texas Sea Grant, 137
Texas General Land Office, 92
    Coastal Division, 107
Texas Marine Mammal Stranding Network, 92
Texas Point National Wildlife Refuge, 159
Texas State Aquarium, 92
U.S. Army Corps of Engineers–Southwestern Division, 109
U.S. PIRG Field Office, 92
University of Texas (Marine Science Institute), 133

Virgin Islands
Buck Island National Wildlife Refuge, 159
Buck Island Reef National Monument, 148
Christiansted National Historic Site, 148
Coastal Management Program, 107
Department of Natural Resources-Division of Coastal Zone
    Management, 107
Department of Planning and Natural Resources (Division of Fish
    and Wildlife), 115
Green Cay National Wildlife Refuge, 159
Island Resources Foundation, 95
national historic sites, 148
national historical parks and preserves, 148
national monuments, 148
national parks and preserves, 95, 148
national wildlife refuges, 159
The Ocean Conservancy, 95
Salt River Bay National Historic Park and Ecological
    Preserve, 148
Sandy Point National Wildlife Refuge, 159
Virgin Islands Conservation Society, 95
Virgin Islands National Park, 148
Virgin Islands National Park and Biosphere Reserve, 95

Virginia
American Sportfishing Association, 92
Back Bay National Wildlife Refuge, 159
Boat U.S., 92
Chesapeake Bay Estuarine Research Reserve, 162
Chincoteague National Wildlife Refuge, 159
Citizens for a Better Eastern Shore, 92
Citizens for the Preservation of Assateague, 92
Coastal Management Program, 107
Coastal Society, 93

College of William and Mary (Virginia Institute of Marine
    Science), 134
Colonial National Historic Park, 148
Cousteau Society, 93
Department of Environmental Quality (Chesapeake Bay and
    Coastal Program), 107
Earth Force, 93
Eastern Shore of Virginia National Wildlife Refuge, 159
estuarine research reserves, 162
Featherstone National Wildlife Refuge, 159
Fisherman Island National Wildlife Refuge, 159
Future Fisherman Foundation, 93
Marine Resources Commission, 115
Mariners Museum, 93
Mason Neck National Wildlife Refuge, 159
Monitor National Marine Sanctuary, 93, 142
National Coalition for Marine Conservation, 93
national historical park, 148
national marine sanctuary, 142
National Maritime Center, 93
National Science Foundation, 113
National Taxpayers Union, 93
national wildlife refuges, 159
The Nature Conservancy, 93
Navy League of the United States, 94
Occoquan National Wildlife Refuge, 159
The Ocean Conservancy–Office of Pollution Prevention and
    Monitoring, 94
Old Dominion University, 134
Plum Tree Island National Wildlife Refuge, 159
Restore America's Estuaries, 94
Riverkeeper–James River Association, 94
Sea Grant Consortium, 138
Sierra Club, 94
Sirenian International, 94
Society for Conservation Biology, 94
Surfrider Foundation–Virginia Beach, 94
U.S. Geological Survey, 111
Underwater Ordnance Recovery, 94
University of Virginia
    Center for Oceans Law and Policy, 134
    Virginia Sea Grant, 138
Virginia Aquarium Marine Science Center, 95
Virginia Beach City Council/Clean the Bay Day Inc., 94
Virginia Eastern Shorekeeper, 95
Wallops Island National Wildlife Refuge, 159

Washington (state)
1000 Friends of Washington, 98
American Cetacean Society–Puget Sound Chapter, 95
Association of Professional Observers, 95
Audubon Society, 95–96
Black Hills Audubon Society, 96
Cascadia Research Collective, 96
Center for Whale Research, 96
Citizens for a Healthy Bay/Commencement Baykeeper, 96
Coastal Management Program, 107
Columbia Riverkeeper, 96
Copalis National Wildlife Refuge, 160
Department of Ecology (Shorelands and Environmental
    Assistance Program), 107
Department of Fish and Wildlife, 115
Dungeness National Wildlife Refuge, 160
Ebey's Landing National Historical Reserve, 148
EPA Region 10 Office, 113
estuarine research reserves, 162
Everett Shorelines Coalition, 96